"This is the best book yet written on social evolution. Jonathan Turner synthesizes his life-work, from cladistics of human great ape ancestors, reconstructing the biological steps that made humans much more emotionally responsive, simultaneously allowing greater brain size and more flexible social arrangements with strangers. Blending symbolic interaction and interaction ritual, early humans developed internalized symbols, self-control, and group references. These let humans build larger, more complex, stratified, and impersonal organization—turning against original individualistic, freedom-loving human nature and submitting it to the social cage. Turner traces the conflict of biological human nature and social organization into post-modern societies and peeks at our future."

Randall Collins, *University of Pennsylvania*

"This remarkable book is both unusually comprehensive and at the same time highly readable. After a slow start, sociology is now being integrated with the findings of evolutionary biology, with Jonathan Turner in the lead. This treatment of human nature and its evolution is powerfully eclectic, using theories and data ranging from primate ethology to theories of emotion to brain science, and includes some pleasant surprises in the form of American Pragmatism and the work of Mead and Cooley. A provocative synthesis."

Christopher Boehm, *Professor of Biological Sciences, University of Southern California*

"Jonathan Turner can be counted among the few in American sociology who ask huge questions, master sprawling literatures, and defy the imperialism of radical social constructivism. He takes nature seriously and wants to know what nature means for humanity. This book continues and extends Turner's decades-long project of systematically understanding and explaining foundational concerns about humanity—that is, *us, we ourselves*. Not everyone will agree with his story, but I commend it as important and fascinating nonetheless. At a time when the authority of science itself is increasingly publicly questioned, Turner admirably models a long-view scholar taking genuinely interdisciplinary science seriously."

Christian Smith, *William R. Kenan, Jr. Professor of Sociology, University of Notre Dame*

"This book by the internationally well-known sociologist, Jonathan Turner, is the one that I personally have been waiting for. Turner is a path-breaking intellectual in evolutionary sociology, neurosociology, and the sociology of emotions. *On Human Nature* is the ultimate summary of his brilliant theory of what made us human. His vision is truly breathtaking!"

Armin W. Geertz, *Professor Emeritus, Aarhus University*

"Jonathan Turner is one of few social theorists who cross disciplinary boundaries in a serious way, engaging biology, anthropology, evolution, genetics, brain science, psychology, and sociology. Rejecting the tautological logic of the 'just-so stories,' so often associated with evolutionary work, Turner reveals the labyrinth-like complexity of human nature. Turner is a sure-footed guide through these labyrinths, rendering his insights useful for thinking about a wide variety of social phenomena. Ultimately, Turner's *On Human Nature* is a cutting-edge work that should matter to all social scientists."

Erika Effler-Summers, *University of Notre Dame*

ON HUMAN NATURE

In this book, Jonathan H. Turner combines sociology, evolutionary biology, cladistic analysis from biology, and comparative neuroanatomy to examine human nature as inherited from common ancestors shared by humans and present-day great apes. Selection pressures altered this inherited legacy for the ancestors of humans—termed *hominins* for being bipedal—and forced greater organization than extant great apes when the *hominins* moved into open-country terrestrial habitats. The effects of these selection pressures increased hominin ancestors' emotional capacities through greater social and group orientation. This shift, in turn, enabled further selection for a larger brain, articulated speech, and culture along the human line. Turner elaborates human nature as a series of overlapping complexes that are the outcome of the inherited legacy of great apes being fed through the transforming effects of a larger brain, speech, and culture. These complexes, he shows, can be understood as the cognitive complex, the psychological complex, the emotions complex, the interaction complex, and the community complex.

Jonathan H. Turner is 38th University Professor of the University of California System; Research Professor, University of California, Santa Barbara; and Distinguished Professor at the University of California, Riverside. He is also Director of the Institute for Theoretical Social Science, Santa Barbara, California. He is the author of hundreds of research articles and the author of more than 40 distinguished books, including most recently *The New Evolutionary Sociology* (with Richard Machalek).

Evolutionary Analysis in the Social Sciences
A series edited by Jonathan H. Turner and Kevin J. McCaffree

This new series is devoted to capturing the full range of scholarship and debate over how best to conduct evolutionary analyses on human behavior, interaction, and social organization. The series will range across social science disciplines and offer new cutting-edge theorizing in sociobiology, evolutionary psychology, stage-modeling, co-evolution, cliodynamics, and evolutionary biology.

Published:

On Human Nature: The Biology and Sociology of What Made Us Human
by *Jonathan H. Turner (2020)*

Mechanistic Criminology
by *K. Ryan Proctor and Richard E. Niemeyer (2019)*

The New Evolutionary Sociology: New and Revitalized Theoretical and Methodological Approaches
by *Jonathan H. Turner and Richard S. Machalek (2018)*

The Emergence and Evolution of Religion: By Means of Natural Selection
by *Jonathan H. Turner, Alexandra Maryanski, Anders Klostergaard Petersen and Armin W. Geertz (2017)*

Forthcoming:

The Evolution of World-Systems
by *Christopher Chase-Dunn*

Why Groups Come Apart: Fusion-Fission Dynamics in Human Societies
by *Kevin McCaffree*

Maps of Microhistory: Models of the Long Run
by *Martin Hewson*

On Human Nature

The Biology and Sociology of What Made Us Human

Jonathan H. Turner

Routledge
Taylor & Francis Group

NEW YORK AND LONDON

First published 2021
by Routledge
52 Vanderbilt Avenue, New York, NY 10017

and by Routledge
2 Park Square, Milton Park, Abingdon, Oxon, OX14 4RN

Routledge is an imprint of the Taylor & Francis Group, an informa business

Library of Congress Cataloging-in-Publication Data
A catalog record for this title has been requested

ISBN: 978-0-367-55648-8 (hbk)
ISBN: 978-0-367-55647-1 (pbk)
ISBN: 978-1-003-09450-0 (ebk)

Typeset in Minion
by MPS Limited, Dehradun

With age, I realize more and more how much I owe to my mentors as an undergraduate and graduate student. With gratitude that I will never be able to express fully, I dedicate this book to:

University of California, Santa Barbara (1961–1965)

Tomatsu Shibutani

Donald R. Cressey

Walter Buckley

Thomas Scheff

Cornell University (1965–1968)

Robin M. Williams Jr.

William Friedland

Charles Ackerman

Leo Meltzer

Donald P. Hayes

Wayne Thompson

Other Books Authored and Edited by Jonathan H. Turner

Authored Books
American Society: Problems of Structure (1972)
Patterns of Social Organization: A Survey of Human Social Institutions (1972)
The Structure of Sociological Theory (1974)
Inequality: Privilege and Poverty in America (1976, with Charles Starnes)
Social Problems in America (1977)
Sociology: Studying the Human System (1978)
Functionalism (1979, with Alexandra Maryanski)
The Emergence of Sociological Theory (1981, with Leonard Beeghley)
Societal Stratification: A Theoretical Analysis (1984)
American Dilemmas: A Sociological Interpretation of Enduring Social Issues (1985, with David Musick)
Herbert Spencer: A Renewed Appreciation (1985)
Oppression: A Socio-History of Black-White Relations in America (1985, with Royce Singleton and David Musick)
Sociology: A Student Handbook (1985)
Sociology: The Science of Human Organization (1986)
A Theory of Social Interaction (1988)
The Impossible Science: An Institutional History of American Sociology (1990, with Stephen Turner)
Classical Sociological Theory: A Positivist's Perspective (1992)
The Social Cage: Human Nature and the Evolution of Society (1992, with Alexandra Maryanski)
Sociology: Concepts and Uses (1993)
Socjologian Amerykansa W. Posukiwaious Tazamosci (1993, with Stephen Turner)
American Ethnicity: A Sociological Analysis of the Dynamics of Discrimination (1994, with Adalberto Aguirre)
Macrodynamics: Toward a Theory on the Organization of Human Populations (1995)
The Institutional Order (1997)
On the Origins of Human Emotions: A Sociological Inquiry into the Evolution of Human Affect (2000)
Face-to-Face: Toward a Sociological Theory of Interpersonal Behavior (2002)
Human Institutions: A New Theory of Societal Evolution (2003)
Incest: The Origins of the Taboo (2005, with Alexandra Maryanski)
The Sociology of Emotions (2005, with Jan E. Stets)
On the Origins of Societies by Means of Natural Selection (2007, with Alexandra Maryanski)

Human Emotions: A Sociological Theory (2008)
Theoretical Principles of Sociology, Volume 1: *Macrodynamics* (2010)
Theoretical Principles of Sociology, Volume 2: *Microdynamics* (2010)
The Problem with Emotions in Societies (2011)
Theoretical Perspectives in Sociology (2012)
Theoretical Principles of Sociology, Volume 3: *Mesodynamics* (2012)
Theoretical Sociology: 1830 to the Present (2012)
Contemporary Sociological Theory (2013)
*Revolt from the Middle: Emotional Stratification and Change in Post-Industrial Societie*s (2015)
The New Evolutionary Sociology: Recent and Revitalized Theoretical and Methodological Approaches (2018, with Richard Machalek)

Edited Books
Strategies for Building Sociological Theory (1979)
Social Theory Today (1987, with Anthony Giddens)
Theory Building in Sociology (1988)
Handbook of Sociological Theory (2001)
Herbert Spencer's The Principles of Sociology (reprint) (2001)
Theory and Research on Human Emotions (2004)
Handbook of the Sociology of Emotions (2007, with Jan E. Stets)
Handbook of Neurosociology (2013, with David Franks)
Handbook of the Sociology of Emotions, Volume 2 (2014, with Jan E. Stets)
Handbook of Evolution and Society (2015, with Richard Machalek and Alexandra Maryanski)

Contents

Illustrations

Tables

Boxes

Preface

For several decades now, I have been exploring the evolution of humans and human societies. In this book, I address a very old topic—*human nature*—but in a way that is different from most such analyses. The fact that humans are evolved great apes, sharing a very high percentage of their genes with contemporary great apes (i.e., orangutans, gorillas, chimpanzees) means that humans and great apes shared common ancestors over the past 16 to 18 million years. These were the ancestors of not only contemporary great apes but also of humans' more immediate lineage of ancestors, termed *hominins* (for being bipedal). From this hominin line would evolve the various subspecies of early humans, such as *Neanderthals, Denisovans Homo sapiens,* and perhaps even others such as *Homo naledi*.

My goal in this book is to isolate, as best as is possible, *the biological nature of humans from the sociocultural nature of humans*. When examining the view that humans have a fundamental biological nature, we need to understand that our biology allows us to create a sociocultural world that we must adjust to. At the same time, it is important to separate, if we can, biology from the sociocultural world of our own creation. Fortunately, two methodologies in biology—*cladistic analysis* and *comparative neuroanatomy*—allow for isolation of evolved human nature from humans' capacity for culture. Culture and human nature are, of course, constantly interacting as they affect human thought, action, and organization, but any claim about "*a* human nature," must consider our evolved biology inherited from our hominin ancestors and the ancestors of present-day great apes.

This is a sociological analysis of the biology of human nature, and this emphasis is important to the story that I have to tell. Humans are, like any life form on earth, the result of biological evolution, driven by the forces of evolution as now conceptualized by biology. From a sociological perspective, however, the most powerful selection pressure on our hominin ancestors was to *become better organized,* or die and become extinct. It is at first difficult to imagine that humans, as evolved apes, are not nearly as social and group oriented as we might think—given the ubiquity, scale, and complexity of our patterns of social organization. But, the story of human evolution begins with the simple insight that the evolving great apes that would eventually become humans were having problems getting sufficiently organized to survive in the predator-ridden open-country habitats of Africa. Great apes and, hence, our common ancestors with them, are *not* highly social, *not* highly organized, *not* oriented to kinship and family, and in general *not* prone to forming stable and permanent groups. Great apes evolved in the forests of Africa, high off the ground in the terminal feeding areas that lacked enough space, food, or structural support to sustain large groupings. As a result, natural selection made great apes more individualistic, more prone to mobility, and less social than their distant cousins, such as all species of monkeys. They were less likely to form stable groupings in a habitat where moving *away* from concentrations of conspecifics was imperative to survival and reproduction. So, how did hominins become more social and better organized, having inherited the biologically based tendencies toward lower sociality, weaker social ties, and few if any permanent groupings? The answer to this question will reveal how natural selection reworked the great ape neuroanatomy of hominins to make them more social, more likely to form stronger ties, and eventually even likely to create stable *groupings*. Humans' evolved nature is, therefore, an outcome of selection on great ape neuroanatomy to make our hominin ancestors increasingly more social and able to form stronger ties and group solidarities. The story is not simple, but it created an animal that only barely survived the African savanna. And yet, this same animal would eventually come to overpopulate earth. In some ways, the story of how humans came to be the dominant species on earth, save perhaps the bacteria and viruses that may kill us off, is more fanciful and fantastic than the movie stories about apes that apparently replaced evil humans. How did such a large animal like a human being, lacking in ability to form strong social ties and groups, ever get so far with seemingly so little for natural selection to work on? The answer to this question is the story of humans' evolved nature that I seek to tell in this book, which in many ways is more interesting than the story in the movies. Moreover, at our biological core, humans are not as decadent as portrayed in the movies, even as we create sociocultural structures that conflict with our basic nature.

My story is not as graphic and engaging as some portrayals of humans' nature, which, as will become evident, I see as more a product of human culture and patterns of social organization than human biology. However, my story is a more accurate history of human evolution and the emergence of a series of what I term *complexes* of behavioral and organization capacities. These complexes revolve around cognitive, emotional, psychological, interactional, and community behavioral and organizational propensities that are, in essence, elaborations of those evident among chimpanzees and other great apes today, and hence, humans' distant hominin ancestors. And they exist because of the sociological imperative for hominins to develop stronger social ties and more stable groupings without, as is often argued, biological drives for hierarchy, family and kin selection, territoriality, religion, warfare, control, and for other need-states that can make for interesting reading and movie-watching but are distorted analyses of humans' real biological nature. Instead, the complexes of humans' evolved biology outlined in these pages just barely allowed our distant ancestors to somehow survive and, then, over the past 12,000 years, to begin constructing mega societies of thousands, then millions, and eventually billions of individuals that now pose threats to other species on the planet and to the ecology of the planet itself. Moreover, these quiet but powerful forces that drive humans have created problems not only for the ecology of the planet but also for the well-being of humans. For, as these complexes have played out in the building up of modern societies, they have, ironically, made it more difficult to fully realize the biology of our very nature.

Jonathan Turner
Santa Barbara and Murrieta, California

Acknowledgment

The research and writing for this book—*On Human Nature: The Biology and Sociology of What Made Us Human*—was sponsored by the Dickson Award, given each year to a professor emeritus of the University of California at Riverside. I am grateful and honored to have received this award for the 2019–2020 academic year.

1
Humans by Nature?

"Human nature is an extraordinarily complex combination of elements, some that are shared with all other species, some that are shared with some others but not all, and still others that are unique and peculiar to *Homo sapiens*. To deny or ignore any part of this rich heritage in constructing social science theories is to deny or ignore elements of the rich world that are altogether fundamental to understanding human life."

Gerhard Lenski, *Ecological-Evolutionary Theory* (2005, p. 46)

In the social sciences today, many remain hesitant to explore the biology of humans. This reluctance contrasts with early scholarly work in virtually all of the social sciences that explicitly or implicitly addressed the issue of humans' fundamental nature. Yet, since the mid-twentieth century, the Standard Social Science Model has gained a strong foothold in some of the social sciences, particularly sociology. This model argues that large brains and expanded capacities for thinking have allowed humans to construct their own world, seemingly somewhat independently from their biology. Spoken language and the ability to symbolically represent any aspect of the universe with culture have led many social scientists, especially in sociology and anthropology, to advocate for a more "constructivist" view of humans' sociocultural world. Over recent decades, sociologists have increasingly viewed humans as unique because of their intelligence, language facility, and capacity to develop and use symbolic culture.

The fact that humans can die, and hence are not "gods," should dispel any notion that humans have transcended their biology or their origins as a life form on earth. Humans emerged from evolutionary processes working on their ancestors' bodies and underlying genome to make us what we are—emotional, smart, and language-using animals that can create and store information. Yet these characteristics hardly liberate humans from their biology. Nor do they mean that human behaviors, thoughts, actions, culture, and social structures are not influenced by their evolved biology. Fortunately, some in the social sciences, particularly economics and psychology but sociology as well, have reengaged with biology. In the case of sociologists, the efforts in economics, psychology, and particularly biology to explain societal-level processes are viewed as inadequate. Today, biologists and social scientists engage in healthy if sometimes contentious debate about the biology of humans and their sociocultural creations, even as many sociologists still proclaim biology to be irrelevant. Some in sociology have gone so far as to argue that exploring biology will rekindle racism, sexism, and a host of other negative "isms," whereas, in fact, biological analysis of human nature will argue for just the opposite. A serious engagement with biology will reduce the dogmatism in the "constructivist turn" in sociology and, moreover, reduce *any* notion that humans are somehow fundamentally different from each other. A better understanding of human biology and how it evolved can remove the pernicious aspects of placing people in socially constructed categories that often contribute to discriminatory treatment.

What follows in the pages of this book is one effort to demonstrate the utility of a biologically informed sociology and to give a new face to a very old topic: What is human nature? Such a question requires that we try to discover *the biological basis of human emotions, thinking, and actions.* As will be evident, it is often difficult to separate what is biologically based from what is determined by the constraints of social structures and their cultures. Still, I offer two approaches borrowed from biology that can help isolate human biology more than many other efforts to describe human nature. Before embarking on this search for the humans by nature in the next chapters, let me review a variety of approaches to illustrate what must be done to isolate, if we can, the biological basis of humans' fundamental nature.

Approaches to Understanding Human Nature

The easiest approach to understanding human nature is one that we have all pursued—whether philosopher, social scientist, or opinionated observer of the "human condition." It is simple speculation about what is biological and what is not. This approach is often not far off the mark because we are all keen observers of what human traits seem to be fundamental and therefore have a biological basis. Another approach, often related to speculation, is what might be called "reverse engineering." It occurs when we examine human behaviors and organizational patterns today and then decide which of these behaviors and organizational patterns are basic to human biology in the past, when societies were not so complex. Although these efforts can be useful, they do not disentangle what is biologically based from the nonbiological structural and cultural constraints on human thought and actions.

Yet another set of approaches comes from biology and psychology. These approaches search for the basic adaptive problems that humans faced during their evolution to distinguish what natural selection wired into the human genome from what is more purely cultural. One technique examines what makes individual humans unique and different, under the presumption that much of this uniqueness is part of our biology. Another technique looks for "universals" or social patterns and supporting behaviors that are evident in all societies, with these universals likely reflecting humans' fundamental biological nature.

I will begin with this last approach, but before doing so, let me just briefly mention here the approach that will guide my analysis, beginning in the next chapter. To discover the biology of human nature, I look at the animals that are closely related to humans genetically—the great apes—and outline those behaviors and organizational propensities they all have in common as likely candidates for the biological basis of humans' evolved nature. Then, I compare the neurology of great apes with that of humans, presuming that differences between the two reflect the work of natural selection as it reworked the basic great ape brain during hominin and human evolution. Doing so makes it possible to analyze the biological inheritance of humans from the ancestors of present-day great apes and, at the same time, discover what natural selection changed as humans' basic biological nature evolved. That said, my goal in this chapter is not to outline in any systematic way the alternatives to my approach, rather it is to provide a sense of how the search for human nature has been conducted in the past and the resulting somewhat chaotic catalogue of humans traits that define what is human nature—at least among those willing to entertain an examination of the biology of human nature. This book could have been titled *The Biology of Human Nature*, but sociology is my focus here. First of all, humans' nature, as distinct from great ape nature, was honed by natural selection under the sociological imperative for humans' hominin (bipedal great apes on or near the human line) ancestors to get better organized or go extinct! And secondly, because the biology of human nature evolved to increase the organizational capacities of hominins, this nature was set up to create societies. In understanding this simple fact, we can begin to see whether various societies since the first hunting and gathering bands are compatible with this nature. At the end of this book, then, we will be in a position to assess the degree to which contemporary societal formations are compatible with humans' evolved biological nature.

Searching for Human Universals

American anthropology was, at one time, much like early American sociology: both looked for what is generic and universal in human behaviors and patterns of social organization. In the early twentieth century, however, American anthropology began to see cultural traits of peoples as unique, refuting the idea of universal patterns in either human behavior or organization. Ethnographies of traditional peoples were treated as descriptions of unique and distinctive cultures rather than as examples of various ways that certain universal behaviors and organizational propensities operated in various environments. In the eyes of these anthropologists, there are no laws of human behavior or social organization because cultures are highly variable.

This view of anthropology dominated much of the twentieth century and, within cultural anthropology, is still dominant. In the early 1990s, however, Donald E. Brown (1991) wrote a detailed review of classic works by leading anthropologists of their time to demonstrate that their ethnographies were empirically flawed and reached the wrong conclusions in highlighting differences among cultures when, in fact, the actual data demonstrate just the opposite: regularities in patterns of behavior and organization. Brown outlined some of the dissenting views against the long-standing rejection of cultural universals among humans. Starting with his own illustrative listing of obvious universals among humans, Brown noted that humans (1) breathe, (2) prepare and eat food, (3) engage in sex to reproduce, (4) produce and use language through phonemes, morphemes, and syntax, (5) perceive the same primary colors in different cultures, (6) create families, religion, economies, polities, and other universal institutional systems, (7) develop divisions of labor, (8) make and use tools, (9) show the same primary emotions in facial gestures, (10) activate the same psychological defense mechanisms to protect self, (11) recognize status differences, and so on.

Even at the height of this cultural relativism in the mid-twentieth century, there were other dissenting voices. For example, A.V. Kidder (1940) pulled together materials highlighting the similarities in patterns of social organization to document the universal properties of all humans and their sociocultural creations. Indeed, he suggested that there might be "specific biological bases for certain of our social habits" (Kidder 1940: 514). Similarly, David Bidney (1947: 391) argued that "cultural phenomena are not intelligible apart from the structure and functions of human nature" (see also Bidney 1944). Turning to some of the most famous anthropologists, Clyde Kluckhohn in his "Universal Categories of

BOX 1.1 PETER MURDOCK'S PARTIAL LIST
OF UNIVERSAL CULTURAL ELEMENTS

Age-grading, athletic sports, bodily adornment, calendar, cleanliness training, community organization, cooking, cooperative behavior, cosmology, courtship, dancing, decorative art, divination, division of labor, dream interpretation, education, eschatology, ethics, ethnobotany etiquette, faith healing, family, feasting, fire making, folklore, food taboos, funeral rites, games, gestures, gift giving, government, greetings, hair styles, hospitality, housing, hygiene, incest taboos, inheritance rules, joking, kin-groups, kinship nomenclature, language, law, luck superstitions, magic, marriage, mealtimes, medicine, modesty concerning natural functions, mourning, music, mythology, numerals, obstetrics, penal sanctions, personal names, property rights, religious rights, propitiation of supernatural beings, population policy, postnatal care, pregnancy usages, residence rules, sexual restrictions, soul concepts, status differentiation, tool making, vising, weaning, and weather control.

Peter Murdock, "The Common Denominator of Culture" (1945: 124)

Note: This is only a partial list of what could be counted as universal human traits to be found in cultures around the world. The question is: What portion of these is the outcome of human nature? What is purely cultural? What involves both humans' biological nature and culture? Whatever the answers to these questions, it is clear that each and every culture is not unique because of human nature and/or because of common responses to organization in diverse habitats.

Culture" (1953) argued that it was a "tautology" to think that only culture determines itself (see also Kluckhohn 1959). Claude Levi-Strauss (1949) sought to explain variations in kinship as a reflection of three universal "mental structures" (recognition of rules, exchange, and bonds between givers and receivers). Functional anthropologists, such as Bronislaw Malinowski (1944), saw variations of human cultures as adaptive responses to limited sets of universal biological, psychological, and organizational needs within their environments. In 1945, George P. Murdock, who, like Malinowski, could be classified as both a sociologist and anthropologist, wrote "The Common Denominator of Cultures," which, because of his creation of the Human Relations Area Files, gave his comments special empirical significance. He then profited a "partial list" of "cultural universals," which is reproduced in Box 1.1 (Murdock 1945: 124).

Despite these few lines of dissention to the cultural relativism dominating American anthropology, scholars such as Glifford Geertz (1965) brought American anthropology back to its relativistic daze, arguing that a universal constant like human biology, psychology, and sociology cannot explain cultural variations (see also his earlier statements [Geertz 1953, 1959], which are more sympathetic to cultural universals). Thus, by the first decade of the twenty-first century, cultural anthropology in the United States had moved away from searches for human universals, whether in humans' biological nature or in their patterns of social organization. Indeed, American cultural anthropology had become not only highly relativistic but predominately anti-science. Thus, the search for universals and the fundamentals of human nature were abandoned. Yet, Brown (1991: 146) lists nine propositions made by critics arguing against the search for human universals. All of these are, I believe like Brown, false; and thus, I am confident that it will be possible to discover humans' biological nature:

1. Human nature and culture are distinct realms.
2. Nature manifests as instincts (seen as fixed action patterns), while culture manifests itself as learning patterns of behavior.
3. Because human nature is the same everywhere, it is culture that explains differences between human populations.
4. Human universals are likely to reflect human nature.
5. Except for its extraordinary capacity to absorb culture, the human mind is largely a blank slate.
6. Therefore, by assertions #3 and #5, culture is the most important determinate of human affairs.
7. Explaining human behavior and organization by human nature rather than culture is a reductionist fallacy; indeed, trying to explain people by anything but culture is reductionist and, hence, not useful.
8. Being an autonomous realm of reality, culture has an arbitrary and highly variable character.
9. Therefore, by #5 and #8, there are few universals that can explain humanity and their creations.

The following argument refutes belief that there are no universal properties of human societies and humans' biological nature. Thus, in reading the preceding arguments against this search for universal features of humans and their sociocultural formations, this book should be considered the case against all these propositions. To get started, let me just emphasize a few points that guide the arguments in this book.

First, there are clearly universal features of human behavior and social organization; and there are many. And the goal of social science, as a science, should be to develop abstract laws explaining the operation of these generic and universal forces creating social structures and culture. Second, while reality is layered into different domains (e.g., physical, biological, behavioral, and sociocultural) and while each of these levels of social reality has its own laws explaining its operative dynamics, the forces of any level can affect the dynamics at other levels. This does not mean that laws on forces are reducible to each other, but it does mean that forces in the physical, biological, and psychological realms can have significant effects on each other and on the universals of the sociocultural universe studied by sociology (and in the past, by anthropology as well). Third, to understand the sociocultural universe, it is useful to determine the effects of humans' evolved biology on human behavior and on human patterns of social organization. Even though these effects will not explain all or perhaps even most universals, they provide a better basis for isolating human nature, or those biology-based behavioral capacities and propensities affect human social organization. Fourth, many universal forms of social organization are not, therefore, easily explained by human nature because they result from human responses to adaptive problems encountered by populations and lead to changes in social structures that are *not* driven by humans' biological nature. Finally, the evolution of forms of human social organization can work *against* humans' biological nature, creating new kinds of selection pressures on societies to change these forms of organization to be more compatible with humans' evolved biology. Knowing the points of strain between humans' evolved nature and the structure and culture of contemporary societal formations is thus useful in tuning societies more compatibly with humans' biological nature.

Darwinian Selection and Biological Analyses

Charles Darwin (1859) along with Alfred Wallace (1858) hit upon the notion that species of life forms are often in competition for resources within niches in various habitats. Members of a given species reveal variations in their physical and behavior traits, although Darwin did not know what caused variations in members of a species (the answer was in Darwin's library, but apparently Gregor Mendel's [1866] famous manual went unread and would not be rediscovered for some time). Members of a given species revealing those traits that enable them to compete for resources will be more likely to survive and produce offspring, whereas those members of a species revealing traits that are unsuited for competition and securing resources will be less likely to survive and reproduce. Later, when Mendel's work was rediscovered, his notion of *merkmals* was converted to the concept of *genes* as the fundamental units of inheritance in all life forms. From this rediscovered knowledge about the units of inheritance governing the structure of phenotypes of life forms, the Modern Synthesis in biology began to take shape and include the following elements:

1. Natural selection works on phenotypes and, thereby, the underlying genotypes of *individual* organisms.
2. Selection favors those phenotypical traits that increase *fitness*, or the ability to survive and reproduce, while selecting out of a population those traits that do not promote fitness.
3. Variations in phenotypes and underlying genotypes are generated by:
   a. distributions of traits, often on bell curves
   b. mutations (breaks in genetic material on chromosomes, creating new alleles or even genes)
   c. genetic drift (selection working on a species in different habitats and niches)
   d. gene flow (as subpopulations of species come into contact via migrations and begin to interbreed)
4. Genes are to be conceptualized at the population level as a *gene pool* of a population of organisms; and even though selection is on individual phenotypes, *it is the population and the distribution of genes in its gene pool that evolves.*

In this view of evolution, an organism's bodily structure and its behavioral propensities are determined by the alleles of genes in the organism's genotype. The problem in much analysis in the Modern Synthesis is how to separate learning and experience (as these affect behavior) from the influence of genes on behaviors. Human nature revolves around genetically driven phenotypical traits governing particular behaviors that have increased fitness of individuals and the population of individual organisms. However, the more a species depends on learning to shape behaviors, the more difficult it becomes to separate genetically driven from acquired behavioral propensities. This problem is accelerated by species like humans, who not only learn and pass on learning across generations but also use language to create cultural systems that regulate conduct as much as the genes in an individual's genotype. Thus, for our purposes in this book, isolating behaviors driven to a significant degree by *an organism's biology,* and particularly its genotype, becomes problematic and often conflated with those cultural forces also directing human behaviors.

Even with the advances in mapping the genome of humans and other animals (and plants), it can still be difficult to separate what is biologically based and what is socioculturally based and influenced by learning and socialization among humans. The result has been a considerable amount of speculation, even as emphasis in biology shifted to understanding the human genotype. Indeed, too often behaviors that seem universal among humans were simply *assumed* to have a strong genetic component and, hence, were part of humans' fundamental nature. These past speculations by more contemporary thinkers were not much better than those among early philosophers, hundreds of years ago, who did not have the benefit of a full understanding of evolutionary dynamics. For example, Konrad Lorenz studied the "aggression instinct," which he saw as channeled by rituals, with natural selection equipping humans' early ancestors with genetically driven mechanisms for using rituals to manage aggression and domination. Other such speculations followed, such as Robert Ardrey's portrayal of the "killing instinct" in *African Genesis* (1961) and *The Territorial Imperative* (1966). Desmond Morris's (1967) views in *The Naked Ape,* along with Lionel Tiger's and Robin Fox's (1971) "imperial animal" and sociologist Pierre van den Berghe's (1973) early analysis of age, sex, and "domination," made rather speculative leaps by assuming that universal features of human societies, per se, made them biologically based and, even worse, that universal features of present-day societies were good indicators (through reverse engineering to their origins) of what is part of human nature.

Thus, "instincts" among humans to exert power, to form families, to be religious, to engage in exchange, to perceive justice, to control territory, to make war, to form groups, and many other seemingly universal human behavioral propensities were assumed to be part of human nature—without much convincing evidence that such behaviors are, in fact, programmed into the human genome. As we will see, when methodologies more adequate to the task of separating biological from sociocultural programmers are used, about half of these assertions are probably *not* true, especially the ones about humans' need to hold power, to dominate, to form families, to construct hierarchies, and the like. Granted, once these are taken off the table as part of human nature, what is left can seem rather tame and less interesting. Nonetheless, much of what has been popularized as part of human nature is, in fact, the result of evolving patterns of social organization and their cultures rather than the result of biological evolution of the human genotype. Human nature is still interesting, though less dramatic, and I hope to demonstrate that such is the case.

Sociobiology and Genic Selection

A significant shift in biology occurred when biologists began to try to explain human behaviors and societies. This shift occurred as a reaction—indeed a rather dramatic overreaction—to early "group selection" arguments (Wynne-Edwards 1962, 1986). C. Wynne-Edwards argued that natural selection often works on individuals to create genetically controlled behavioral propensities to form groups as a "strategy" to increase the fitness of each individual and its ability to pass on its genes. Although obvious to a sociologist, this concept was not so obvious to biologists already locked into a set of biases about the nature of evolution. Even more surprising, the notion of group selection was threatening to biologists committed to the Modern Synthesis. As a result, some began creating a new field, termed *sociobiology*, devoted to explaining the biological basis of social organization *without* resorting to notions of group selection, even the rather tame one postulated by Wynne-Edwards.

Calling this response an overreaction by sociobiologists is based on Wynne-Edwards's (1962, 1986) apparent argument that selection was not working on the group, per se, but rather *on individuals' phenotypes* and underlying genotypes and that selecting on prosocial and altruistic behavioral propensities could lead to group formations that enhanced all individuals in the group. Nothing in Wynne-Edwards's argument goes against the dogma of the Modern Synthesis that selection is on the individual, but it is the population and its gene pool that evolves. Thus, Wynne-Edwards's approach did not appear to challenge any of the dogma of the Modern Synthesis in biology, but such was not how group selectionist arguments were perceived. Thus, *sociobiology* was created to describe how group behaviors could be explained by selection on only individual behavioral propensities. In this assertion, the behavioral propensities discovered were seen as lodged in human biology and under genetic control and, hence, could rightly be seen as part of human nature and as an explanation that would transcend sociology—obviously a contention that most sociologists rejected.

What apparently bothered early sociobiologists was any implication that selection was working *on the group* or, more generally, patterns of social organization of a species rather than on the *individual* phenotypes and the underlying genotypes of *each individual* in a population. Selection *on* the group rather than on individuals would not be Darwinian selection, and it would change the entire model of biology from a view of selection working *only* on the individual and underlying genotype of each individual leading to the evolution of a population of genotypes or gene pool. Moreover, group selection would suggest that not only is the group *directly being selected upon* but also that the group and pattern of social organization (not the genes in the gene pool) are evolving—a further break from the assertions of the Modern Synthesis. Such an argument would not be heresy to most sociologists, but to biologists and to sociologists who adopted the assumptions of sociobiology, it was heresy. It would violate the quasi-religious commitment to the Modern Synthesis as holy scripture handed down from early saints such as Charles Darwin, although Darwin himself was never as dogmatic as his followers (see Darwin 1871 [1875], 1872). For to even imply a divergence from the dictates of the Modern Synthesis was blasphemy; if social structures and their cultures are the units on which selection works, as well as the units that are actually evolving, strictly biological explanations of human nature and patterns of social organization would not be relevant to sociology. Such a conclusion would run counter to the goal of sociobiology in E. O. Wilson's (1975) original formulation proposing sociology as a subfield of biology, an assertion that was hardly welcomed by most in the sociological community and one that Wilson has fortunately withdrawn.

As sociobiology sought to discredit the notion of group selection, it shifted focus from phenotype to gene, thus leading George C. Williams (1966) to posit the notion of *genic selection*. Phenotypes were just the "carrying case" or "temporary homes" for genes, which are what actually evolves. Behaviors controlled by genes are what determined the fitness, or the ability to survive and reproduce. Thus, it was *genic selection* (natural selection working on genes by selecting on the phenotypes housing them) that drove evolution, thus shifting the focus of biological explanations to genes. Richard Dawkins's book *The Selfish Gene* (1976) provided an engaging metaphor that viewed genes as seeking out "survivor machines" in their blind search for immortality. Genes do not die like bodies do; they are passed along across generations of bodies under the guidance of natural selection. Those genes that increase fitness of phenotypes and behaviors will be carried along in successive fitness-enhancing survivor machines that keep the genes in a population's gene pool. Yet, Dawkins did recognize that humans possessed a new kind of replicator and regulator for creating new forms of survival machines—*memes*—organized into cultures (rather than genome). These memes would allow humans to build new and more flexible survival machines—that is, *sociocultural formations* as survival machines. With this admission, then, it became once again difficult to separate the power of genes from the power of memes; and as a result, it became problematic to isolate human nature housed in a biological phenotype from the nature of humans' other survivor machine, sociocultural phenotypes.

Nonetheless, sociobiology continued to offer explanations that directly focused on *human nature*. The first big ideas of how individual behavioral propensities are driven by genes to stay in the gene pool were arguments for *kin selection*. Humans will naturally form relations with relatives with whom they share genes, and the more genes that they share with others, the more they will seek to protect them. Thus, family and kinship systems are not so much born of altruism but by selfish genes seeking to remain in the gene pool by creating useful survival machines—families and kinship systems from which more complex survivor machines could potentially be built. When the frequency of non-kin helping each other was pointed out, the notion of *reciprocal altruism* developed, arguing that genes push individuals to exchange favors with those who can help maintain their genes in the gene pool (Trivers 1971, 2005). Again, genes seeking to survive in the gene pool are what drive humans to help each other; acts of one individual are reciprocated by another, thus keeping both of their respective genes in the gene pool. Sociologists, such as Van den Berghe (1972, 1973, 1975, 1981), picked up on these ideas and saw them as means for building patterns of cooperation, or societal organization, driven by genes seeking survival machines. He extended sociobiological arguments in a number of ways.

For reciprocal altruism, he argued that reciprocity may not be returned because it is more fitness enhancing to receive benefits from others without reciprocating. Van den Berghe then argued that this reality favored larger brains and cognitive capacities to detect and punish *free-riders* and *cheaters* who fail to reciprocate. He also recognized that exchanges are limited in how well they can organize larger populations, suggesting that *coercion* and *use of power* are alternative mechanisms that enhance fitness. In other words, selection favored genes that pushed for coercion to coordinate larger numbers of individuals and for larger brains to produce cultural ideologies *legitimating power use*. It is not clear why selection must do this at the genic level, however. Could not selection be working on smaller groupings such as bands? As they got larger, social control problems led to sociocultural selection on the band as a whole to invent a system of politics and its legitimation in culture. Is the biological explanation any better? For someone committed to finding so much in human biology, the biological explanation seems better. In fact, however, the biological argument put forth by Van den Berghe is no more convincing than the sociological argument emphasizing selection on sociocultural formations rather than individual phenotypes. Thus, separating genes from memes is not so easy, and thus, separating what is in humans' biological nature and what is built into a culture-using species like humans is not easily achieved, except by fiat and just-so stories, which are the stock and trade of much sociobiology.

Van den Berghe (1981) also sought to explain ethnic dynamics in terms of an extended view of kin selection. Historically, larger kin groups (lineages and clans, for example) constituted a breeding population of close and distant kin who would sustain trust and solidarity with each other while mistrusting more distant populations. He coined the term *ethny* for ethnic groupings, which he viewed as extensions of more primordial kin groups containing individuals who shared certain cultural traits (and perhaps physical traits as well) that increased bonding and solidarities (overlapping with kin relations through intermarriage) as a fitness-enhancing strategy to protect ethny members. Thus, while Van den Berghe's argument is much more complex than those in socio-biology, its basic thrust is based in sociobiology: genes push to find survivor machines—in this case an ethnic subpopulation—where co-operation and solidarity ensure member genes remain in the gene pool.

Other sociologists who adopted this meaning of sociobiology developed additional explanations. For example, Arlen Carey and Joseph Lopreato (1995) argued that genic selection drives efforts of human populations to cut reproduction rates when resources are scarce, thus increasing the likelihood that genes will not be selected out of the population by keeping its members roughly in equilibrium with available resources. Individuals may feel that this is prudent and rational, but it is actually their genes seeking immortality that drive such thoughts (see also Lopreato 1984, 2001; Lopreato and Crippin 1999).

Thus, sociobiology is a theoretical approach that would seem to be compatible with any effort to discover the biology of human nature. But, as we will see, some of its arguments simply do not hold up, especially the notion of *kin selection*. Others, however, such as *exchange reciprocity*, do seem to have a biological basis. These extreme assertions of socio-biology and the willingness to ignore how culture also influences the very processes viewed by sociobiologists as driven by genic selection prevent us from jumping too rapidly into this mode of explanation. There are, as I will try to demonstrate, better alternatives.

Evolutionary Psychology

Evolutionary psychology retains most of the basic tenets of sociobiology but inserts an emphasis on the evolution of the human brain. Early evolutionary psychology (e.g., Cosmides 1989; Cosmides and Tooby 1992; Barkow, Cosmides, and Tooby 1992) emphasized that the human brain was composed of *modules* or specialized bundles of neurons that had evolved in response to selection pressures during the late Pleistocene when hominins (bipedal ancestors of humans 4 to 1 million years ago) were evolving. This line of emphasis followed Wilson's mandate of sociobiology as "the systematic study of the biological basis of all social behavior" (1975: 4). More recently, notions of modules have been relaxed and given less emphasis because they do not quite fit the way the brain has actually evolved over the past few million years. More specifically, theorizing and research in evolutionary psychology has focused on *behaviors* as evolved adaptations in response to adaptive problems faced by those species on the human evolutionary clade (Alcock 2001: 10–16, 23–40). Unlike sociobiologists, evolutionary psychologists were less concerned with the ultimate causes of a behavioral adaptation than with the adaptation of humans and their immediate ancestors. Among the first topics studied by evolutionary psychologists (Symons 1979) were *sexual strategies* and *mate selection* (Buss 2016: 287; Kenrick, Maner, and Li 2016: 930–933). This emphasis is within Darwinian tradition because of its implication for how humans have achieved such reproductive success and, hence, fitness to pass on their genes.

Borrowing from Robert Trivers's (1972, 1974) ideas on *parental investment* and *sexual selection*, both mates are seen to derive the same benefit from reproduction, since 50% of each's genes are passed on to the next generation and remain in the gene pool. These concepts contribute to prolonged, even lifetime, joint parental care of offspring because of their respective investments in their offspring. In the hands of evolutionary psychologists, however, diverse mating strategies ranging from lifelong monogamy to variants of polygamy, including polygyny, polyandry, and promiscuity, can evolve among men and women (Buss and Schmitt 1993). Men gain much more in fitness (passing their genes on) in short-term relationships than do women because women inevitably have a longer-term investment in one offspring than a male. Also a female's lifetime reproductive success is limited by the time required for the fertilization of her limited number of eggs and the prolonged infant and juvenile care in the societies where mating strategies first evolved. In contrast, males can at little cost (and perhaps great pleasure) impregnate larger numbers of women and, thereby, increase the number of babies that will carry his 50% of the offsprings' genes. Still, males and females in studies conducted by Buss and colleagues (Buss 1989; Buss et al. 1990) identify the same top four qualities of mates among both men and women: (1) mutual attraction and love, (2) dependable character, (3) emotional stability and maturity, and (4) pleasing disposition (see also Hopcroft 2016: 35). It is likely that males and females will still reveal somewhat different evolved psychologies because females seek out partners with resources to ensure survival of her genes, such as higher social status, possession of economic resources, prospects for financial success, age, maturity, evidence of ambition, and behavioral stability and dependability (Buss 2016: 104, 5–14), coupled with physical cues associated with masculinity. The evolved psychology of females is more complicated than this quick review, but it differs from the males' because a male can have many more offspring than can females if he is promiscuous. Men will prefer traits marking female health and fertility often idealized by standards of beauty and perhaps sexual fidelity on the part of females. Again, the evolved psychology is more complex, but the basic argument is clear (see Turner and Machalek 2018: 216–234 for a longer summary of evolved psychology of mate selection and retention). This line of argument supposedly explains why men are more jealous than women and why men will be more violent against women when they perceive female infidelity.

Sociologists have also adopted the basic approach of evolutionary psychology and have generally tried to explain *rates of particular kinds of behavior* in terms of evolved psychologies, such as high male rates of crime and especially violent crime over low female rates, higher rates of male promiscuity compared to females, higher rates of incest between fathers and daughters (especially stepdaughters) compared to females' rates of incest with sons, and so on. The effort to isolate modules in the brain continues but is less essential now than looking at different behavioral syndromes of individuals generated, ultimately, by the brain. The documentation of differences is typically accurate, but the evolutionary explanation about how these differences reflect the biology of the evolved psychology behind behaviors is less developed and is often argued as a just-so story (e.g., Donald 1991, 2001). In many cases, equally persuasive arguments can be developed on how patterns of behavior are influenced more by culture than biology, thus again conflating the search for what is basic in human nature and, in the case of evolutionary psychology, what is basic to female versus male nature. As will be evident, the alternative approach I take in examining the evolving psychology of humans is to look back much further in time at the biology of the animals from which the human line evolved. Rather than examining rates of behavior today, where conflation of culture and biology is difficult to avoid, it is wise to examine behaviors and organizational patterns *before* there were humans and their cultures in order to see what might have been passed down the evolutionary clade leading to humans.

Some More Purely Sociological Approaches

Gerhard Lenski's Evolutionary-Ecological Approach

Gerhard Lenski was one of the first sociologists to reintroduce evolutionary analysis back into sociology in the second half of the twentieth century (Lenski 1964; Lenski and Nolan 2005) and into the twenty-first century (Lenski 2005; Nolan and Lenski 2018). In his last major work, Lenski (2005) began by asking what humans share with all other species, as a form of life. Humans are organisms composed of one or more cells built from the same basic materials: minerals, water, organic compounds, fats, proteins, and nucleotides and their derivatives (Lenski 2005: 34). We evidence a metabolism that revolves around nutrition, respiration, and synthesis on which survival as a life form depends. This metabolism can be converted into energy to act in the environment in order to secure resources necessary for sustaining and reproducing life. All forms of life contain genetic codes in all cells that, with latitude to environmental conditions, are the blueprints for life forms. These points are all pretty obvious, but they bring home a fundamental assertion: Humans *are* biological creatures that are programmed to some degree because we are built from the same basic ingredients as all forms of life. Relatedly, humans are the product of natural selection as it worked on the phenotypes—bodies and behavioral propensities—and the underlying genotypes for these phenotypes. This simple fact should also emphasize that we *cannot ignore biology-based propensities* directed by genotypes.

Lenski then narrowed the comparison among species to what humans share with certain classes of species rather than all life forms. One of the most important is that humans are *organized into societies*. Another is that *learning is an important source of information*, supplementing genetic codes. Still another is that we depend on *active communication* among conspecifics. And yet another is that humans evidence *an individual identity* vis-à-vis other conspecifics and pursue *both* self-interest *and* collective interests. Although some might think that these traits begin to liberate humans from their biology and the work of natural selection on the human genome, they should, instead, emphasize that these traits *are the outcome of natural selection* on human biology. Yet, there is a sociology in these as well, and it is this combination of biology and sociology that often makes trying to isolate just the biological part difficult.

Finally, Lenski turned to what makes humans unique or, at least, distinctive. His list included *bipedalism* or standing upright (other animals can do so, but not habitually), *tool making* (again, other animals do make and use tools but not to the degree of humans), *communicating* (beyond biology-based calls with cultural symbols), *using language* and communication to create *vast stores of culture* (beyond what we know other animals can do). Lenski argued that human nature is distinctive because humans have *fundamental needs* for (1) sex, experiences, affirmations of self, (2) securing resources to satisfy needs beyond what is necessary for survival, (3) maximizing pleasures and minimizing pain, (4) derivative needs from experiences and cultural programming, (5) individual differences that cannot be ignored in social relations, (6) verifying a sense of self by others, and (7) pursuit of self-interests combined with altruism for the collective. Each of these supposedly unique capacities and behaviors may not be wholly unique to humans, and so, Lenski moved into speculation in his views about human uniqueness. More likely, humans are somewhat unique because we can do all these *to a greater degree* than other animals. Still, I will try to specify more clearly whether they are unique to humans and, more specifically, if they are part of humans' evolved nature. Obviously, some, like the need for sex, is not unique to humans, but as we will see, sex among humans has been a critical force behind the evolution of culturally regulated patterns of human social organization.

Speculations in Sociological and Related Social Science "Theories"

Within sociology, each of the major theoretical traditions posits something fundamental about humans. For example, exchange theories and all their variants argue that humans are motivated and driven to pursue rewards or utilities and to pursue those lines of conduct where they can make, if possible, a "profit" (that is, rewards exceed the costs and investments to get these rewards [e.g., Blau 1964; Coleman 1990; Hechter 1987]). Behaviorism makes many of the same assumptions (Homans 1962, 1974; Emerson 1962), without the calculation of costs and rewards but emphasizing that individuals are motivated to gain rewards and that behaviors that have been rewarding in the past will elicit the behaviors that allow an organism, including humans, to get this reward now and in the future. Utilitarian arguments in both economic neoclassical theory and in rational choice theories in sociology sometimes make a more extreme assertion: it is in human nature to try to *maximize utilities* (rewards) by calculating costs and accumulated costs (investments) in order to gain utility.

Symbolic interactionist theories of all varietals emphasize that the driving force of humans is the effort to verify or affirm, in the eyes of others, self-conceptions and identities of a person as a certain type of individual, deserving of certain responses from others (Burke and Stets 2009;

Stets and Serpe 2016; Turner 2002, although many only see efforts as a behavioral propensity rather than a need-state). Thus, human nature revolves around the behavioral propensity to present to others in all situations various types of identities and, conversely, to read the gestures of others in order to understand the identities that these others are presenting.

Ritual theories in sociology argue that individuals are oriented to arousing positive emotions allowing for feelings of solidarity with others (e.g., Durkheim 1912; Goffman 1959, 1967; Collins 1975, 2004). Sometimes, a notion of self is introduced into these theories, but it is a somewhat different view compared to how symbolic interactionists conceptualize self. In ritual theorizing, if self is added to the theory, individuals strive to present self in a certain manner and to get others to accept this presentation of self. In contrast to symbolic interaction, which is more likely to emphasize self as a series of more enduring identities, the self presented to others as conceptualized by ritual theory is *situational and often transitory*. That is, what motivates individuals is to present a self, whether genuine or not, and have others accept this self for the duration of interaction. Thus, for ritual theorists, the notion of codified identities is overdrawn, whereas symbolic interactionists would view ritual theorizing as capturing only a small part of human nature.

More phenomenological theories and philosophies emphasize the need to achieve a sense of meaning. Individuals are motivated to "make sense" of what is transpiring, to see events as forming understandable patterns and coherent meanings and thereby allowing a person to gain a sense of what some have termed *ontological security* (Giddens 1984; Schutz 1932). Without this sense that the broader context and situational interaction in this context "are as they seem," individuals experience anxiety and, thus, become even more motivated to achieve a sense of "facticity" (Garfinkel 1967; Turner 1987, 1988, 2002) as well as other psychological states in which persons feel they understand a situation.

Conflict theories, especially those derived from Karl Marx, all emphasize that humans have certain fundamental needs. Marx, for example, emphasized that humans need to have control of their labor to the extent that they determine what productive activity they will participate in, how they will engage in this activity, and to whom they will distribute the products of their labor. This assumption about humans' fundamental nature reinforced Marx's view that capitalism took the ability to realize these needs away from humans, thereby alienating them and, eventually, leading them to mobilize for revolt against their oppression and exploiters. More contemporary critical and conflict theories will generally argue more broadly, assuming that humans have fundamental needs to control their own destiny and to experience a sense of efficacy in making decisions about how they will act. All cultural and social structure constraints that deny this sense of efficacy are exploitive and repressive, violating human nature.

In various versions of gestalt theorizing, the primary need of persons is to experience balance and consistency in social relations and in their cognitions about social relations. Social relations, for example, are out of "balance" when relations with others are not transitive, as is the case of not liking the friend of a friend. In such situations, individuals like their friend and do not like the friend of their friend, creating a lack of transitivity that then motivates the person to seek balance by either rejecting the friend or embracing the previously disliked friend of a friend. At a more cognitive level, gestalt theories emphasized individuals' attempts to keep their cognitions consistent; when there is cognitive dissonance, one of the cognitions will have to change. Thus, in a situation such as unbalanced relations among friends, the individual holds two dissonant cognitions (I like my friend; I do not like the friend of my friend). This dissonance in cognitions can only be resolved by deciding to defriend the friend or to befriend the friend (of my friend), thus restoring cognitive consistency. Many other theories and philosophies posit this notion that humans are motivated to maintain balance and harmony at various levels of thought and behavior. Individuals thus seek harmony with their social and physical environment, consistency in their cognitions of any importance, and balance in their social relations.

Many philosophical traditions have either created more scientific theories or the reverse. For example, pragmatist philosophy was a distinctive philosophical orientation born in the United States in the nineteenth century. The central idea was that humans are always trying to adapt to their environments. Therefore, because the principal environment for humans is other humans organized into patterns of sociocultural organization, humans are driven to exhibit those behaviors and actions that increase their adaptation to ongoing social relations and patterns of social organization. Symbolic interactionism is one theoretical tradition in sociology that emerged out of pragmatist philosophy, arguing that the unique behavioral capacities of humans are to possess and present a sense of self to others, to role-take and read the gestures of others in order to determine their disposition to act and their evaluations of a person's self-presentation, and to make minded decisions over potential options for behavior in a situation. These "learned behaviors" facilitate adaptation. At the same time, later pragmatists emphasized that these behavioral capacities were only made possible by the neurology of humans as it evolved through natural selection in creating humans as a distinctive species.

Another theoretical approach to understanding human nature arose out of the early philosophizing of contract theories (e.g., Hobbes 1651; Locke 1689; Rousseau 1762) and then early economics to produce utilitarianism (Smith 1776) and the neoclassical school of economic thought (e.g., Olson 1967 [1971]). Humans are motivated to reduce what today are sometimes called "negative externalities" or harmful things, such as conflict, tensions, uncertainty, and so on, through the willingness to write a social contract with a source of power that will regulate these negative things in return for granting legitimacy to that source of power. This contract tradition evolved into neoclassical economics, as espoused by Adam Smith (1776) and later into a variety of rational choice and exchange theories within all the social sciences (Olson 1967 [1971]; Hechter 1987; Coleman 1990). The basic need of humans is to gain utilities and, at times, to give away certain freedoms in order to regulate and control negative outcomes, thereby maximizing or gaining utilities under less than ideal conditions. Thus, humans are, by their nature, willing to cede control to centers of power and to cultural beliefs/norms/ideologies to reduce negative externalities in order to gain more utilities, with the ceding of some control as simply a cost of seeking particular types of utilities.

Christian Smith's Analysis of "Personhood"

Christian Smith's (2010) book *What Makes a Person?* represents yet another type of sociological approach to human nature. Smith's book is, in essence, an ontology on the nature of human beings for, as he notes, "human beings have an identifiable nature that is rooted in the natural world" (Smith 2010: 10); and the goal is to determine "what is actually real about a human being." With humans, *the person* represents layers of higher reality built up from adaptations to interactions with the environment. Personhood develops as a kind of layering of capacities, with those at the top being built upon the foundations of those lower in the hierarchy. Moreover, a constant flow of influence occurs up and down the hierarchy. Table 1.1 lists the 30 capacities that Smith sees as essential to being a person. For the present, reading from the bottom up represents a kind of evolutionary development of humans (Smith would probably be hesitant to phrase it this way, but it suits the arguments that I develop in later chapters). This set of capacities, beginning with "existence capacities" and moving through "primary experience capacities" up to "secondary

TABLE 1.1 Christian Smith's List of Fundamental Human Capacities

Complexity of Capacity	Specific Human Capacity
Highest-order capacities	30. Interpersonal communication and love
	29. Aesthetic judgment and enjoyment
	28. Forming virtues
	27. Moral awareness and judgment
	26. Truth seeking
	25. Abstract reasoning
Creating capacities	24. Self-reflexivity
	23. Truth seeking
	22. Anticipate the future
	21. Valuation
	20. Compose and recount narratives
	19. Language use
	18. Symbolization
	17. Create, grasp, and communicate meanings
	16. Self-transcendence
	15. Material cultivation
	14. Inventing and employing technology
	13. Creativity, innovation, and imagination
	12. Acting as efficient causes of own actions
Secondary experience capacities	11. Intersubjective understanding
	10. Episodic and long-term remembering
	9. Emotional experience
	8. Interest formation
	7. Assigning causal attributions
	6. Practical consciousness
	5. Volition
	4. Mental representations
	3. Understand quantity, quality, time, and space properties
Existence capacities	2. Conscious awareness
	1. Subconscious awareness

Source: Christian Smith, *What Is a Person?* (2010: 54).

experience capacities" and, then, to "creating capacities" and, finally, to "highest-order capacities," highlights *elaborations* in which one set of capacities can expand as a result of the growth of the hominin brain, the evolution of spoken language, and the capacity to accumulate and transmit culture. Smith is less concerned with the biological basis of these elaborated capacities; rather, his concern is with the fact that humans possess these capacities. Indeed, they are *what makes for a person.*

This long and provocative book is, therefore, not concerned directly with human nature as I am framing the issue: What is the biological basis of human capacities and behaviors that are used to build up social structures and their cultures? Rather, his concern is with understanding the nature of being human in its broadest sense and what this means for how sociological inquiry should be conducted. What I find most interesting about Smith's list in Table 1.1 is how it overlaps with many of the capacities I discuss in later chapters, although my concern is how these capacities are tied to human biology and, in particular, human neurology. Capacity #9 is the key, because without the capacity to be emotional, the brain cannot grow; and if the brain cannot grow, language (#19) and later speech cannot emerge; and if language and especially speech cannot emerge, culture in the human measure cannot exist, thus reducing all other capacities listed from #20 to #30 in the hierarchy.

I would essentially agree with Smith's assertion that "the thirty causal capacities are the stuff out of which human personhood exists emergently," but my goal is more evolutionary: How and why did these capacities emerge, and what is the biological basis of each, if any? In contrast, Smith is more interested in mapping the paths of emergence and then, once a capacity emerges, its reverse causal effects on the very capacities that led to its emergence. In the end, he constructs a complicated map of causal arrows linking the capacities. While I will not discuss each of the 30 capacities listed by Smith, nor will I try to construct a complicated mapping of their causal connections, I will focus on those capacities that evolved by means of Darwinian natural selection and were then elaborated upon as the biology of species on the human evolutionary clade evolved. Many of these capacities are evident in other primates, especially other great apes, that are closely related to humans genetically. Yet, my evolutionary analysis will present a picture—indeed, a kind of causal map—of human nature that is not too far from the analytical scheme that Smith has developed, although much simpler. That is, as capacities emerge, they feed back and affect the capacities from which they emerged—a process of *elaborations*. My goal is to add a better sense of which ones are *more central to humans' biology-based nature* built upon the capacities of humans' distant primate ancestors. Thus, I think it useful to go back in evolutionary time—millions of years ago—to see how these capacities that constitute the essence of humans have evolved. Only in this way can we see which are fundamental and how later capacities were built up from those inherited from hominins on the human clade of evolution.

Nicholas A. Christakis's "Social Suite" of Universals Necessary for Societies

Just as I finished a first draft of this book, another rather intriguing book titled *Blueprint: The Evolution Origins of a Good Society* (2019) was published. Its author, Nicholas Christakis, was trained as both a medical doctor (MD) and as a sociologist (PhD), and he runs a large research operation that is both broad and fascinating. One of his basic goals has been to highlight, on the basis of a wide variety of data, that there is a "social suite" of universal behaviors and organizational propensities among humans that are universal to the species and, when executed in a particular way, present a "blueprint" for societies that meet fundamental human needs. Naturally, given what I try to do in this book, this approach was fascinating, although his goal is somewhat different from mine: I am trying to isolate the biological basis of human social organization as inherited from humans' last common ancestors with present-day great apes. For, as will become evident, humans' biological nature is an elaborated version of great apes' biological nature, which should not be surprising because humans share an ancestor with the ancestors of each of the extant great apes that still survive today. Indeed, humans and great apes are closely related, in a genetic sense. My goal is to separate what is biology based from what is a product of sociocultural activity by humans, whereas Christakis seeks to discover the right mix of behavioral and organizational attributes for creating and sustaining "a good society." His findings are quite compatible with mine, but he leaves a bit unclear the distinction between what is driven by our genes and what is constructed by humans' capacities for thinking, speaking, and using culture to organize social life.

The social suite consists of:

- the capacity to have and recognize individual identity
- love for partners and offspring
- friendship
- cooperation
- preference for one's own group ("in-group bias")
- mild hierarchy (relative egalitarianism)
- social learning

Although rather general, these attributes are nonetheless fundamental. Yet, Christakis illustrates with many quite fascinating examples how getting one or two of these wrong makes for a society that does not meet human needs, and indeed, makes for a society that will disintegrate.

Instead of providing a "social suite," my analysis eventually leads to the formulation of a series of "complexes" of behavioral and organizational traits that can be considered biology based. These are termed the *evolved cognitive complex*, the *evolved psychology complex*, the *evolved emotions complex*, the *evolved interaction complex*, and the *evolved community complex*. In some ways, my analysis can provide detail about the origins of these biology-based traits that are clearly part of humans' biological nature. They came from humans' distant hominin ancestors over the past 5–6 million years with the split of hominins from the ancestors of contemporary chimpanzees. Previously, 8–9 million years ago and even earlier (13–16 million years ago) the common ancestors of present-day gorillas and orangutans split off from the line that would eventually lead to hominins and then humans. Each of these elements in Christakis's social suite is a product of what natural selection was, first, doing to the great ape line over 30 million years of evolution in the arboreal habitats of Africa and, then, to the human hominin line as the ancestors of humans had to increasingly adapt to more terrestrial, open-country habitats.

Having to leave the forests of Africa as they receded over 10 million years (due to episodic cooling of the forests in Africa) placed enormous selection pressures on humans' hominin ancestors to get better organized or go extinct. Christakis's "social suite" is one outcome of these selection pressures. As will become evident, however, my portrayal of the evolution of the five *complexes* of human nature undergirds all the elements in Christakis's social suite and, I think, provides some clarification about how this suite evolved. Bioprogrammers for nuclear family, pair-bonding among adults, tight-knit groups among cooperating individuals, and even adult friendships were *not* prominent in the great ape genome, despite Christakis's argument. Indeed, it was *the lack of some of these elements* in the social suite of early hominins carrying the great ape genetic legacy that represented the biggest roadblock to the emergence of humans.

My story is, in many ways, a description of the routes that evolution took to bring about this social suite. The fact that humans still have difficulty in getting the elements of this suite together in the right proportions—as is highlighted by Christakis in his many illuminating examples—signals that being "social" may not be as "natural" to humans as is often assumed. So, Christakis's and others' efforts to discover human universals as a window to human nature is not as straightforward as it seems. Much of what is universal in humans is similar behavioral and organizational responses to similar adaptive problems of social organization. Such universals are, of course, outcomes of humans' inherited biology as it is mobilized to create new sociocultural formations in response to environmental selection pressures. But are these universal responses *directly governed* by the human genome? Or are processes more complex? I think the second question, emphasizing the complexity of this process, is closer to what I am seeking: the biological basis of human behavior and social organization. This book seeks to discover humans' behavioral and organizational capacities that were inherited from the common ancestors of great apes and hominins and, then, elaborated upon by natural selection as it worked on the neuroanatomy of hominins. The social suite and many other hypothesized universals are not *direct expressions* or manifestations of biology-based drives or bioprogrammers but, rather, outcomes of humans' evolved neurology as it has been mobilized to create new sociocultural formations in the face of common adaptive challenges. For much of what makes humans "social" is built upon what made *great apes less social.* Human sociality is based, as contradictory as this may seem, on a genetic foundation of low-sociality species that did not form families, permanent groups, pair-bonds, and even friendships to the degree necessary to complete the social suite. This remarkable achievement of how natural selection worked to make hominins and then humans more social and more group oriented was not the result of mutations to the brain, nor installations of new bioprogrammers, but of other changes to the brain caused by directional selection on existing neurostructures of hominins that would allow humans to *create* new forms of culture and social structure to solve similar adaptive problems. Social structure and culture are not driven by human biology; rather, humans' neurological capacities allow humans to *construct and* create *what is not always natural,* in a biological sense, to an evolved ape. And one of these creations is the social suite, which has some biological basis but not to the degree that some would argue. The reason that the social suite is so difficult to get right resides in the fact that it has to be created and re-created in constant adjustment to the copresence of others in changing environments. Humans have to *work at* creating and sustaining a viable social suite because it is not a set of traits directly governed by bioprogrammers; rather, the social suite is achieved by a neurology that enables humans to be highly emotional and intelligent and to speak and, thereby, use their hardwired capacity for language to create symbolic culture.

What Is Human Nature vs. What Are the Outcomes of This Nature?

Human evolution has both biological and sociological elements, as implied by the subtitle to this book. On the one hand, humans are an evolved great ape and, hence, an animal constrained by its genome. On the other hand, humans are a special animal because *Homo sapiens* possess the ability to build up social structures and cultures that are *not* genetically regulated but instead are regulated by *nonbiological* forces. It is in the nature of humans to create sociocultural formations, but these formations are not human nature itself; they are the *outcome* of the biological nature of humans as its suite of capacities expanded over many million years of evolution.

This duality of humans—what is part of humans' biological evolution, and what is part of human action and thinking generated by sociocultural evolution—is often difficult to separate and sort out. Nonetheless, we attempt it because sociocultural formations can often violate humans' biological nature, thereby placing humans in conflict with their biological nature. There is, moreover, always a feedforward and feedback relation between humans' biological nature and their sociocultural constructions. Humans can build social structures and their cultures because of their biological capacities as they make for human agency. Once constructed, however, these formations can constrain, violate, or enhance humans' biological nature. I will argue that the first societies constructed by humans—bands of nuclear families engaged in hunting and gathering—were the outcome of selection forces working on individuals' phenotypes and underlying genotypes to make hominins (bipedal ancestors of humans) and then humans more social and better organized. This simple structure and its culture were the "survivor machines" of humans *for most of their evolutionary history*, which only began to change over the past 12,000 years as populations settled down, grew, and, as a consequence, were forced to create new and more complex organizational forms in order to survive. Most of these new forms of social organization *violated basic features of human nature*, but they persisted, nonetheless, because they were essential to human survival (Maryanski and Turner 1992).

Yet, as I noted earlier, when social structures significantly thwart humans' biological nature, there emerges a persistent tension that operates as a kind of "selection pressure" to reorganize societies to better fit human nature. Later evolution of industrial and postindustrial forms of social organization are at best 300 years old, and hence, quite nascent in an evolutionary time scale. These new forms are not as compatible with nature as the first societies of hunting and gathering, which reflected humans' evolving biology under pressures to get better organized. They are, however, more compatible with this nature than all other societal types since hunting and gathering. Thus, once we have a clearer picture of what the biology of human nature involves, we can assess sociocultural formations by the degree to which they inhibit nature, or conversely, allow for expression of this nature.

Conclusion

As will become evident, the pages to follow seek to separate humans' biological nature as it was inherited from the common ancestors of humans' closest living relatives, the great apes. Humans are, in essence, evolved great apes; and not surprisingly, much of what we are at the biological level is a great ape with a larger brain, with the capacity to speak and to build sociocultural formations. The biology behind all these capacities is revealed in human neurology. And so, another way of isolating out humans' biological nature is to engage in comparative neuroanatomy in which specific structures of great ape brains are compared to human brains, with the differences representing what natural selection did to the basic great ape brain *to make us human*.

In this first chapter, I have touched on various ways that scholars in different disciplines have tried to determine what is fundamental to being human. And in so doing, the human capacities touched upon in the text, lists, and tables of this chapter only offer a rough sense for the range of ideas about human nature. Taken all together, I doubt if human nature is this complex, but elements of a good many of those features of being human that have been mentioned for illustrative purposes here will emerge as central to humans' biological nature. At the same time, it is important to recognize that much of the sociocultural universe is constructed to meet selection pressures for adaptation of populations of individuals and has not always reflected human nature. There is, as we shall see, less of a one-to-one relationship between a biological capacity and the evolution of specific behaviors, social relations, cultural systems, and social structures. Despite some relationship, other forces are almost always in play in the human sociocultural universe. Nonetheless, I still think that we will be able to focus in on the most fundamental aspect of human nature. That nature has allowed humans to create mega or macro societies and their cultures that reveal their own internal dynamics. A primary goal here, then, is to be able to make a reasonable judgment about the extent to which particular patterns of behavior and social organization among humans have a biological basis, and equally important, which do not. And not only these alternatives but also which behaviors and patterns of sociocultural organization work against expression of our nature as humans. For we are indeed "humans by nature," but our ability to express this nature is often thwarted by a sociocultural universe of our own making.

2

Before Humans
Looking Back in Evolutionary Time

Humans are evolved great apes, and much of our basic nature can be discovered by comparing ourselves to great apes. This comparison is often done by merely studying behaviors of great apes and comparing these to humans' behaviors. Such comparisons, while interesting and useful, are limited because it is also apparent that great apes act and organize differently from humans. Despite what movies have proclaimed, there can be no "planet of the apes" because apes cannot organize like humans can. And so, direct comparisons of present-day great apes with humans are less useful for my purposes: *to uncover the biological basis of human nature inherited from the ancestors of present-day great apes and humans.* Fortunately, data on the behavior and organization of great apes, and primates more generally, can be used for my purposes with a methodological procedure biologists call *cladistic analysis.*

Cladistics shifts our orientation to *looking back in time* to the common ancestors that humans' ancestors once shared with the ancestors of extant great apes. How is this done when biology lacks the equivalent of the Hubbell telescope that can look back to almost the beginning of the physical universe? Cladistics allows us to see what is not directly observable: our last common ancestor with great apes as well as the evolution of our more direct and immediate ancestors—termed *hominins*—who split off from the ancestors of contemporary chimpanzees some 5–6 million years ago (mya).

The Power of Cladistic Analysis

Cladistic analysis makes the assumption that species that are closely related are descendants of a common ancestor—what can be called the *last common ancestor* (LCA). Once the genetic relatedness of a set of species is established, it is possible to tabulate the traits they all share and also some traits that only some share. The more traits are common among all the species under study, the more reasonable it is to assume that the LCA possessed these traits (Andrews and Martin 1987; Kitching et al. 1992[1998]; Maryanski and Turner 1992; McGrew 2010). The logic of cladistics is similar to what is done in historical linguistics, in which a set of languages thought to have evolved from a root or "mother tongue" are examined for what they have in common. It is then assumed that what is common was part of the original root language (Maas 1958; Jeffers and Lehiste 1979).

In cladistic analysis, two essential hypotheses constitute basic assumptions: One is the *relatedness hypothesis*, which holds that similarities in traits of what are assumed to be related species are not due to chance but, in fact, are the outcome of descent from a common ancestor. Thus, if traits in the three great apes—orangutans, gorillas, and chimpanzees—are similar, it is assumed that they are the outcome of having come from the common ancestors of these great apes. The second assumption in cladistic analysis, the *regularity hypothesis*, holds that modifications from the ancestral form to the descendant forms are not randomly acquired but reveal a clear systematic bias that links these descendants to each other and to their LCA (see Maryanski and Turner 1992: 7–32; Turner and Maryanski 2008: 28–57). For example, if the features of great apes differ from other primates, such as monkeys, it is assumed that this difference of great apes and monkeys is systematically generated and sustained by different selection pressures on species of apes and monkeys (Andrews 2019).

To assess both hypotheses, cladistic analysis typically invokes a quasi "control group" consisting of a sister lineage to the one being examined—in our case, monkeys as a control group for assessing the traits of great apes. Old-world monkeys of Africa are a different line of evolution from the great apes and, thus, can serve as this control group for the analysis of the common traits of great apes and their LCAs. Great apes and old-world monkeys both adapted to the arboreal habitat, but species of monkeys diverged from apes about 25 to 20 mya. Because of the different niches in the arboreal habitat that they occupied, the species' divergence initiated a systematic bias in the selection pressures working on apes and monkeys. Thus, we can be confident if the traits of apes and monkeys differ that this difference is due to the evolution of great apes in a different set of niches from monkeys and, moreover, that this divergence is the result of extant great apes having a common ancestor from which each evolved. Otherwise, one would have to assume that the common traits of great apes evolved independently of each other—an unlikely event. However, traits that great apes do not share can be seen as the result of selection as it worked on the neuroanatomy as each of the sister species (chimpanzees, gorillas, and orangutans) evolved in somewhat different habitats and niches since breaking off from their LCAs. For example, if chimpanzees reveal differences in certain key traits from orangutans and gorillas, these differences are the result of selection on chimpanzees in ecological niches that differ somewhat from those of gorillas and orangutans.

With these assumptions and a control group provided by old-world monkeys, we are now in a position to see the results of cladistic analysis on the three species of present-day great apes: orangutans, gorillas, and chimpanzees. In so doing, we also catch a view into the distant past of humans' ancestors because *Homo sapiens* and all other closely related species that emerge on the threshold of being human—such as *Homo neanderthalis*, *Denisovans*, and *Homo naledi*—were all descendants of the ancestors of contemporary great apes via the hominin line that separates humans from the ancestors of chimpanzees some 5–6 mya. One way to visualize common ancestors is with a cladogram as presented in Figure 2.1.

Perhaps it is difficult to see hairy great apes, and even those with red hair (orangutans), as related to humans, or worse, humans as just another great ape but with less hair all over their bodies. Yet, with the sequencing of the genomes of all the great apes, including humans as the fourth extant great ape, it is clear that great apes and humans are closely related. As might be expected from the cladogram in Figure 2.1, orangutans are the least related to humans because they split off from the common lineage that humans have with other great apes some 12–18 mya, which is reflected in the fact that orangutans share "only" 96–97% of their genes with humans, albeit with their similar genes spread out on an extra chromosome pair as is also the case with gorillas and chimpanzees as well. Next are gorillas, who share about 98% of their genes with humans, again spread across 48 instead of the 46 chromosomes for humans. And finally, humans' closest living relative, chimpanzees, share 99% of their genes with humans. There can be no doubt that humans and the great apes are closely related, which, of course, makes the great apes a good subject for cladistic analysis and a surprisingly clear vision of the distant past, as much as 16–18 mya. We can see what the last common ancestor was like, one that was much like the present-day orangutans that remain in the forests of Asia. You can see on the cover of this book a plate of the three great apes standing next to a human, with their hair cut off. All great apes can easily stand up, although the cover plates adjusts their feet and hands in a somewhat unnatural position. The natural locomotion pattern of great apes is a kind of hopping action with hand knuckles pushing off the ground, but the ability of great apes to easily stand, suggests that the conversion of hominins to bipeds was not as dramatic was it might seem. What is even more fascinating is that when great apes are raised with humans, they often take on a human identity;

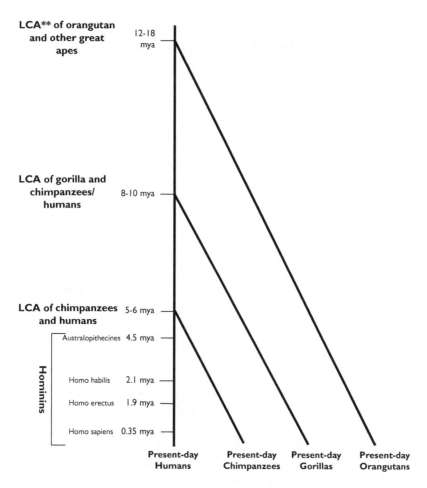

Figure 2.1 Cladogram Depicting Splits of Common Ancestors of Humans with Great Apes

Notes:
* Humans are considered great-apes, granted very evolved great apes
**Last Common Ancestor of present day orangutans with the ancestors of other greatapes on the clade, and so on for the last common ancestors of gorillas with chimpanzees and humans and, then, the last common ancestors of chimpanzees with hominins leading over 5 million years to humans

and, in some cases, walk around in an upright mode that is not normal when living in the forest. Indeed, data from great apes raised with humans reveals that, when such is the case, they see themselves as human when asked. For example, a great ape that is asked to sort pictures of apes and humans has little trouble doing so, putting pictures of apes in one pile and pictures of humans in another pile. When this ape who has been raised with humans comes to his or her own picture, this picture will generally be placed into the pile with humans. For example, Vickie was a chimpanzee raised in a home; she clearly recognized herself in a mirror and, what is more, declared herself to be human. She learned in her human home to sort snapshots into human versus other animal piles, and she consistently placed her own picture on the human pile (Gallup 1970, 1979, 1982; Gallup et al. 2014). Koko, the famous gorilla raised by her trainer, Penny Patterson, similarly sorted her snapshot into the human pile. A female great ape raised with humans and viewed herself as human showed no interest in ape males when she reached puberty; instead, she only had eyes for young human males. Thus, great apes have a capacity once thought to be unique to humans: the capacity to see themselves as an object, to develop an identity of an alien species (humans), and to align this sense of self to the expectations of others and social situations. (It is now clear that elephants, dolphins, and probably whales also have this capacity, as may also be true for highly intelligent birds.)

If normal biological criteria were employed, chimpanzees would be in the genus *Homo* with humans, but it is difficult for humans to accept that they are so closely related to animals covered in hair and "walking" with feet and knuckles. True, humans are much more intelligent and, as we will see, more emotional than great apes and all other animals on earth; still, humans descended from ancestors that we shared with the present-day apes. And, what is most important, these other great apes *will be able to tell us a great deal about human nature* because, at the level of our biology and neurology, humans are still very much a great ape. Even though enhanced emotions and a brain three times the size of other great apes alters human nature somewhat, we remain at our core great apes. Cladistic analysis thus gives us the tools to see not only our origins but also our essence as biological life forms.

Years ago, Alexandra Maryanski (1986, 1987, 1992, 1993, 1995) conducted the first thorough cladistic analysis of great apes, using species of monkeys as a control group. Her goal was to assess the social ties evident in great ape communities. When coupled with other field data collected on primates over the last 70 years, Maryanski's findings reveal a somewhat surprising pattern. Her approach was to review all the extant, at the time, field studies on great apes and a representative sample of species of monkeys as a control group. Most research in the field is conducted by those interested in behavior, whereas Maryanski was interest in *relationships among individuals* and the social structures created and sustained by relations among age and sex classes in great ape communities. She thus used accounts of behaviors to determine the strength and nature of *social ties among conspecifics* and then to see whether relationships led to *network ties* that could be used as a proxy for determining the nature of the structures organizing great apes and the control group of monkeys. The rather surprising results revealed that great apes do not evidence many strong social ties. Instead, most social ties are weak and episodic, suggesting a weak-tie structure among rather individualistic primates moving about in large communities or home ranges that can span many square miles. It immediately became evident when doing an assessment of the strength of social ties that great apes do not form permanent groups and, most interestingly, not even groups based on kinship ties beyond the universal tie of mammalian mothers to their offspring. Thus, none of the great apes reveal anything close to the nuclear family of mother, father, and offspring because great apes are highly promiscuous, making fatherhood impossible to determine. Moreover, all great apes reveal a pattern of female transfer to a new community at puberty, leaving their mother (and unknown father) forever. As a result, it is impossible to have lineal, across-generational kinship relationships, even partial ones such as a mother and her offspring, when all females depart their community and are replaced by females immigrating from other communities. Moreover, it also became clear that immigrating females evidenced weak ties built around mutual tolerance given to strangers from different communities and sitting in proximity to let their young offspring play. Among orangutans and gorillas, young males at puberty also leave their natal community or at least their mother's location in this community, thus forever breaking up intergenerational family ties. However, chimpanzee males do not leave their community and remain in their natal community for their lifetime, forming cordial relations with their mothers (mostly periodic visits for an afternoon or day), but they do not form a *stable kinship group* with their mothers. Chimpanzee brothers also evidence social relations, but males typically have closer relationships with non-kin male friends. Yet, males and females in all great ape communities are free to move about, alone or in short-term parties, and in general reveal a great deal of individualism and mobility around the territory occupied by their community.

Thus, humans are descendants of animals that did not form (a) many strong social ties, (b) stable kinship units beyond the temporary ties of mother and her young offspring, or (c) stable groupings. In fact, the only stable social unit appears to be a broader sense of the boundaries of a common community, or home range, and a cognitive mapping of who belongs and does not belong to this community or home range. Among chimpanzees at least, they will form temporary patrols to make sure that males from other communities do not enter their home range (by hurting or, at times, killing males who make incursions into their community). It appears that the other two great apes—gorillas and orangutans—also have this sense of the more inclusive community, but it is not clear how far they go in defending its boundaries. Nor is it clear that communities have the well-understood boundaries of chimpanzee communities.

In Tables 2.1 and 2.2, Maryanski constructs a hypothetical portrayal of the social ties and organization of the last common ancestor (LCA) to all the great apes and humans. Table 2.1 outlines the strength of social ties among the species of great apes, with the far-right column representing the cladistic reconstruction of the LCA, or last common ancestor to all present-day great apes and, by the logic of cladistics, humans and their line of hominin ancestors as well. In Table 2.1, *adult-to-adult ties, adult-to-adult offspring ties*, and *adult-to-preadolescent offspring ties* are summarized for gorillas (column 1), chimpanzees (column 2), orangutans (column 3), and their last common ancestor (column 4, far right). A "0" denotes nonexistent or weak social ties, a "0/+" denotes ties of moderate strength, and a "+" points to a strong social tie.

Based on the data, *adult-to-adult ties* are generally weak. Among orangutans and chimpanzees, males and females do not form permanent bonds and evidence only a few moderate ties. These ties include male and female chimpanzees who seem to be friends and, as a result, hang together more than is normal in most chimpanzee communities. Among orangutans, a male may "court" a female for a week or so and at times remain with her for a longer period of time (perhaps several weeks) if danger is perceived. Among gorillas, females with offspring (typically from promiscuous sex with males) will stay around the lead silverback male while her offspring are young, often using him as a babysitter for her independent activities (which include sexual relations with younger males who hang around the fluid group associated with a lead silverback), but she will often leave the grouping when her offspring, at puberty, initiates transfer away from its mother to other communities or to distant locales of another existing community. Among chimpanzees, there are generally moderate ties between brothers, although, as noted earlier, males generally prefer chosen friends among chimpanzees. There is also the moderate-to-close ties of chimpanzee males to their mothers, whom they periodically visit, as is emphasized in the next row in Table 2.1 outlining *adult-to-adult offspring procreation ties*. Otherwise, there are no strong ties among adults in great ape communities.[1] The potentially moderate or even stronger ties of male chimpanzees with males friends

TABLE 2.1 Strength of Social Ties among Extant Species of Great Apes

	Species of Ape			
	Gorillas (Gorilla)	Chimpanzees (Pan)	Orangutans (Pongo)	Last Common Ancestor
Adult-to-Adult Ties:				
Male-male	0	0/+	0	0*
Female-female	0	0	0	0*
Male-female	0/+	0	0	0*
Adult-to-Adult Offspring Procreation Ties:				
Mother-daughter	0	0	0	0*
Father-daughter	0	0	0	0*
Mother-son	0	+	0	0*
Father-son	0	0	0	0*
Adult-to-Preadolescent Offspring Ties:				
Mother-daughter	+	+	+	+*
Father-daughter	0	0	0	0*
Mother-son	+	+	+	+*
Father-son	0	0	0	0*

Notes:
0 = no or very weak ties
0/+ = weak to moderate ties
+ = strong ties
* is used to denote a reconstructed social structure, in this case the likely structure of the last common ancestor to humans and extant great apes.
As is evident, this structure is most like that of contemporary orangutans.

TABLE 2.2 Strength of Ties among Sample of Well-Studied Monkeys

	Gelada	Patas	Macaque (Most Species)	Baboons (Most Species)
Adult-to-Adult Ties				
Male-male	o	o	o	o
Female-female	+	+	+	+
Male-female	o	o	o/+	o/+
Adult-to-Child Ties				
Mother-daughter	+	+	+	+
Mother-son	o	o	o	o
Father-daughter	o	o	o	o
Father-son	o	o	o	o
Adult-to-Adult Ties				
Mother-daughter	+	+	+	+
Mother-son	o	o	o	o
Father-daughter	o	o	o	o
Father-son	o	o	o	o

Notes:
o = weak or null ties
+ = strong ties
o/+ = weak or moderate ties.

and/or their brothers, as well as the moderate ties of females with children to the lead silverbacks, can be assumed to be traits that evolved after gorillas and chimpanzees split off the line of their LCAs, probably under selection pressures imposed on communities of gorillas and chimpanzees in their more terrestrial niches in the arboreal habitat that differed from orangutans, who spend most of their time up in the trees.

The next block of ties, *adult-to-adult offspring procreation ties*, reveals that fathers are not strongly tied to their adult offspring, which would be expected given that paternity is never known because of sexual promiscuity among males and females. Only chimpanzee males evidence a stronger tie to their mothers, which, as noted already, is maintained by cordial visits but not by actually living in a group with the mother. For orangutans and gorillas, strong ties between adult males and their mothers are not possible because they will have transferred away from their mother at puberty. The same is true for all great ape females, who, at puberty, leave their natal community or, at the very least, move to locations within their community that cut off ties with their mothers, thus breaking the strong tie of preadolescent offspring with their mothers (again, fathers are never known because of promiscuity). Thus, with the tie of male chimpanzees and their mothers being the only exception, all females (and males as well) among orangutans and gorillas leave their parents' sphere at puberty, thus breaking the possibility of lineal kinship ties into adulthood. This breaking of ties thus eliminates for gorillas and orangutans *any* cross-generational ties among adults.

The last block of ties, *adult-to-preadolescent offspring ties*, reveals the strongest ties in great ape communities. These mother-son and mother-daughter ties, which are nearly universal among mammals, are broken at puberty for orangutans and gorillas when male and female transfer away from their mothers to another community or to new locales away from their mothers within their existing community. Among chimpanzees, only the daughters leave the community (the sons remain in their natal community for their lifetime).

I am perhaps belaboring these points about social ties and networks, but they are the key to understanding human nature. By just looking at the large number of "0" signs in Table 2.1, then, one gets an immediate sense of the weak-tie world of great apes, with the only moderate-to-strong ties occurring with chimpanzee males with mothers, brothers, and male friends and with universal mother-preadolescent nurturance until puberty, when all females and all males, except for chimpanzees, leave their natal community. Moreover, among orangutans and gorillas, this departure from mothers may not always involve movement to a new community but simply transferring to new locations where boundaries are not as clear as they are among chimpanzees.

Using these data on age and sex classes leads to the cladistic reconstruction of the LCA to all great apes, hominins, and humans in the far-right column of Table 2.1 where the *only* consistently strong tie is between mothers and their preadolescent offspring. It should not be surprising that orangutans come close to matching the LCA, since they were the first to break off (see Figure 2.1) about 12–18 mya and have always lived in the forests of Asia (now mostly limited to Borneo and Sumatra). The result is a pattern of weak ties, no group structures, and no family structures beyond the temporary adult female–young offspring tie. Only with later splits from the LCA by the ancestors of gorillas and chimpanzees, where more time was spent on the ground under the forest canopy, did natural selection begin to wire in a few more strong ties, but even then, this selection *did not* produce permanent groups or kin groups of any stability.

The somewhat stronger ties among lead silverback males and females with offspring as well as the formation of the rather loose and fluctuating group surrounding the lead silverback are the results of selection on these large animals in their semiterrestrial habitat within the highland and lowland forest habitats of Africa. Yet, as noted, the female usually breaks this tie as her offspring transfer away; and while the group congregating around the lead silverback remains, its members are constantly shifting as individuals move about a community or home range. Chimpanzees, who are closest to humans genetically, evidence a much more pronounced evolution of structure, with males staying in their natal community and forming friendships with other males. Yet, *none* of these stronger social ties leads to the formation of stable group structures within the community; again, groups form and disband, parties on community patrol and parties of mobile members of the community form and disband, and many individuals wander around the community alone, making episodic contact with others before moving on.

What is evident, then, is that humans' hominin ancestors descended from animals revealing few strong ties and no stable group formations within what appears to be a larger and, also, a more stable community structure. Thus, community is the only stable structure among the ancestors of hominins on the human line. Humans are thus descendants of highly promiscuous animals that did not form families or any permanent groupings beyond the mother–young offspring dyad and that evidenced individualism, autonomy, and mobility in their social relations within a larger home range or community. Humans' biological nature is, to a great degree, formed by this weak-tie, low-sociality profile of our distant ancestors. But humans are an endpoint on a clade that evolved for 5–6 million years after the split from the LCA that hominins shared with the ancestors of present-day chimpanzees; and so, selection could alter the behavioral propensities, the weak ties, and the low-density network structures created by weak-tie relations. Still, at our genetic core, it is not likely that the ape in us disappeared; rather, it is more likely that new behavioral propensities *were layered over* the more ancient ways of behaving and organizing outlined in the far-right column (with a*) of Table 2.1.

Monkeys and Great Apes

We are now in a position to determine whether the reconstruction of the LCA in Table 2.1 meets the conditions of the relatedness and regularity hypotheses discussed at the beginning of this chapter. Genetic analysis suggests that humans and all the great apes are closely related genetically, and this genetic closeness signals that the ancestors of humans and the great apes constitute a regular and continuous pattern for millions of years. To make this conclusion even more "conclusive," it is useful, as noted earlier, to invoke a control group of species that are distantly related but that, at the same time, constitute a different set of species evolving in a different set of niches. Monkeys and great apes split off from each other around 28–25 mya, although no firm date is known. Apes initially dominated the arboreal habitat as primates evolved but, then, by 20 mya, monkeys had come to dominate the arboreal habitat in Africa. Indeed, today, there are more than 130 species of old-world and new-world monkeys and only four species of great apes, counting humans as the fourth great ape. So, in terms of evolutionary success, great apes have been in decline, except for humans (on whom the verdict is still out); and all other great apes may be extinct in their natural habitats within a century, if not sooner. Thus, great apes are an evolutionary failure, except for humans (who may or may not be on the way to killing themselves off directly or indirectly through destruction of their environment). Even prosimians or pre-monkeys have 62 species, and they are marginal in arboreal habitats dominated by monkeys (Maryanski and Turner 1992: 26–27; Turner and Maryanski 2008: 48).

The absolute number of ties shown here is not as important as whether strong ties can serve as a building block for extended networks and social structures that organize a species. As is evident, the mother-daughter bond is the focal point in monkey societies for building up large, tight-knit networks of female kin in matrifocal cliques, which provide the only basis for intergenerational relationships in monkey societies. Weak-tie relations among males ensure that there is little structure, beyond the competitive, contested, and changing male hierarchy, in monkey societies.

Table 2.2 compares four representative and well-known species of old-world terrestrial monkeys. This table may initially look similar to Table 2.1 (because of the many "0" signs or weak-tie relations), but a large difference is evident in the specific relations. Among monkeys, the strong ties produce a different pattern of social organization compared to great apes. Monkeys, organized at the group rather than the community level, reveal two bases for group organization: one is dense networks of related females who *never* leave their natal community (just the opposite of great apes, where *all* females leave their natal origins), and the other base of social structure is dominance hierarchies among male monkeys who have migrated into a group from another group (thus, in monkey societies, males leave their natal groups, as is the case for orangutans and gorillas but not chimpanzees). The female matrilines consist of collateral *and* lineal relations among females across several generations, whereas the dominance hierarchy emerges and then is reshuffled by competition among immigrating males. Despite seeking to horde females as a harem, dominant males are rarely fully successful, which means that fatherhood cannot be assured. Thus, it is the mother–daughter adult tie that forms the core of monkey groupings; and this pattern is just the opposite of great apes. The organization of monkeys into groups of related females and dominance hierarchies among males is also different from the organization of great apes, again consisting of larger communities with no stable groupings with this community among adults.

These fundamental differences from great ape societies indicate that monkeys and great apes constitute two different clades and that each meets the relatedness and regularity hypotheses. Such organizational differences have allowed monkeys to live in open-country habitats and, in contrast, prevented great apes, except those on the hominin line, from moving out to open-country niches away from the forests. It is *this lack of organization at the group level* that posed the problem for great apes, and it is, no doubt, the reason for their decline.

Why Are Monkeys and Apes So Organizationally Different?

When monkeys began to take control of the arboreal habitat some 20 mya, they were able to control the core areas of the trees, where food is more plentiful, where the space can support larger groupings, and where greater structural support is found in the thicker branches of the tree. One hypothesis for why monkeys were able to take control from generally larger and smarter great apes is that selection gave them the advantage of being able to eat unripe fruit, which allowed them to get at the fruit earlier than great apes and thus take over those niches where it is most abundant. They congregated in larger groups that revealed considerable structure. In contrast, the ancestors of great apes today were forced to the terminal feeding areas high up in the forest canopy, where there is less food, less room, and less structural support by thin branches far from the heavy and thick portions of the tree trunks. As a result, selection worked to ensure that densities at any place the forest canopy would remain low, with both male and female adolescents moving away from their mothers and, because of male-female sexual promiscuity, from their unknown fathers. Whether these distant ancestors had a community organization is less relevant than what they did not have: the female matrilines and male dominance hierarchies that would make for dense organization *at the group level*. This lack of cohesive and stable groups among great apes today

is old, probably at least 20–25 million years old. Without genetically en-
dowed bioprogrammers for groups and close kin relations in the genome
of the ancestors of great apes, selection would face a large hurdle *if* more
permanent groups and kin relations became necessary for adapting to
a more terrestrial habitat where slow great apes would encounter fast,
four-legged predators or packs of predators.

How could natural selection put back what had been taken away, or
that which was never present along the ape line? Humans represent the
answer to this question, and so, if we begin with the features of extant
great apes, we can get a relatively clear picture of what humans inherited from the great ape clade; and these inherited traits will be part
of "human nature." But, clearly, selection had to do something more to
this inherited nature from great apes because none of the great apes
can do what late hominins and early humans did: live in kin groups
and bands in open-country bushlands and savanna conditions. We can
see from Table 2.1 that some additional structure emerged as gorillas
and chimpanzees adapted to more terrestrial niches in the forest habitat. However, because none of the great apes, except one isolated
population,[2] can live out on the savanna, much more structure was
necessary for hominins to be fit for the open country. So, what "new"
traits did natural selection add? And, are these new traits also part of
"human nature"?

The Liabilities of Being a Great Ape

Hominins had many disadvantages as formerly tree-dwelling animals
that were forced to the ground as the forests in Africa receded and, in
their place, came more open-country habitats of secondary forests,
bushlands, and savanna. First, as hominins came to the ground, they
were labelled *hominins* because they became increasingly bipedal,
walking around on two rather than four legs. Yet, even when bipedal,
hominins would be slow compared to four-legged predators that might
feast upon them. Second, hominins were visually dominant, with smell
subordinated to vision. Being visually dominant in the forest canopy is
highly adaptive, but on the forest floor and then out into more open-
country bushlands, grasslands, and savanna, olfaction is more adaptive
because it is easier to smell predators (hiding, perhaps, behind bushes
or still far away) than to see them. Olfaction does not require the same
directional focus as vision, because chemical smells simply diffuse into
an area and automatically alert individuals and groups to potential
danger. Third, great apes and presumably hominins have a tendency to

become emotional and noisy when confronting danger—screeching and otherwise making noise, which, of course, would only attract more predators. Fourth, ape individualism, and being a loud individualist at that, would make hominins even more vulnerable and, being slow as well, easily picked off by predators. Thus, as physical specimens, hominins were not particularly imposing when on the ground in open country. As a result, many open-country great apes probably became easy prey and went extinct. In fact, one somewhat successful strategy of a terrestrial ape was to become enormous and physically imposing, as was the case with *Gigantopithecus*, who grew to 8 feet tall and weighed many hundreds of pounds. They lasted for a long time in more open-country habitats, because of their size; but they also had disadvantages of being slow bipeds, although they may have been quadrupedal but still not very fast or quick. Moreover, they could not live in larger groups because of their need to eat so much food each day. And, they apparently had a hard time expiring heat from their large bodies in warmer climates. So, size alone was not a winning strategy; and because they are not nearly as imposing as was Gigantopithecus is probably one reason gorillas have not ventured out of the protection of the woodlands.

Even more problematic for descendants of great apes was their lack of organizational structures. The only stable structure inherited from the LCA was a sense of community, which, as we will see, had some advantages when human societies began to grow. That said, communities too large and too spread out are not able to coordinate defense and food gathering in the open country. Early hominins, such as *Australopithecines,* were upright but still rather small (4 feet at best) and thus fairly unimposing to predators; and so, they may have initially gone out and then returned to the forests with the first signs of danger. The later more robust forms of *Australopithecines* may have stayed out for longer periods in open country, not only because of their somewhat more robust size but also because they may have become more organized for defense and food collection.

Thus, somehow, selection began to find a way to increase the organization of hominins into more stable and cohesive groups. Most descendants of great apes that went out onto the savanna did not survive, but perhaps late *Australopithecines* and, then, the first true forms of *Homo—Homo habilis* and later *Homo erectus or ergaster*—were increasingly better organized and could safely spend more time in open country. We know that this was the key to their survival for 2-plus million years and to their eventual evolution into the various subspecies of humanlike forms that populated Africa, Europe, and perhaps Asia by 450,000 to 300,000 years ago.

As *Homo erectus* radiated out of Africa to Europe and Asia, it is likely that selection somehow pushed them to become more organized. Otherwise, they could not have easily adjusted to the diverse habitats in the many areas where they could survive. How, then, did these evolving great apes along the hominin line begin to form more stable bands of nomads, and perhaps even a kinship system, built around the nuclear family of mother, father, and offspring? This was the challenge facing natural selection; and for most species of great apes, it appears—though the fossil evidence is scant—that natural selection failed and did not find an answer to the organizational weaknesses of great ape social structures. For hominins, however, natural selection did blindly hit upon some answers, generating some new traits and enhancing some existing ones that became part of humans' evolved nature.

Preadaptations and Behavioral Capacities

The typical answer to the question, What makes humans unique?, is big brains, spoken language, and culture. However, these impressive attributes among humans as a species evolved fairly late along the hominin line. For example, the brain of a chimpanzee, our best proxy for the LCA at 5–6 mya is about 375–400 cubic centimeters (cc); the brain of *Australopithecines* was not much bigger, beginning at 4.5–5.0 mya right up to 1.4 mya when *Australopithecines* died out. *Homo habilis* (1.6–2.5 mya) had, by measurements of fossilized cranial caps, a brain size of around 500 cc, just 20% larger than chimpanzees today. Measurements of *Homo erectus* (1.8 mya) to the first humans vary enormously, from 500 cc to the lower end of the human measure at 1,050 cc (the human brain is, on average, around 1,350 cc). Some of this variation may come from deme effects when *Homo erectus* radiated broadly out of Africa. As a result, subpopulations were probably isolated as a *deme* from others living in somewhat different habitats that pushed for larger brains. The same is true of recent finds in southern Africa, such as *Homo naledi* and others (.05 to 2.1 mya), which are remarkably humanlike in their skeletal forms but with brain sizes not much bigger than those of early *Homo erectus* at 500–600 cc. Thus, there were many species and subspecies of *Homo* stretched across several continents where the brains evolved or remained static without apparent selection pressures for dramatically more intelligence.

Thus, to the extent that spoken language and culture are related to intelligence, which, in turn, is a consequence of brain size, it is clear that none of these were as important as might be thought in resolving the organizational problems of hominins. Clearly, something was happening with *Australopithecines*, but it was not just overall brain growth, which remained at 400–500 cc for several million years. The brain only grew dramatically, I would hypothesize, from 700,000 years ago to the emergence of *Neanderthals* and the first *Homo sapiens* around 450,000 to 300,000 years ago (see Holloway, 2015, for summaries of measurements of fossils). Something caused this growth, and it was probably not the imperative to get better organized because group solidarities are not so much related to intelligence or even spoken language and culture as to something else that is often not adequately addressed: emotions. This increase in emotional capacities of hominins and then humans compared to other primates that increased social ties and patterns of group-level organization. But, as we will see in the next chapter, the early growth that occurred in the hominin brain was not in the neocortex as much as it was in the more ancient *subcortical areas* below and inside the neocortex. Indeed, the modest increase in the size of the brain during the evolution of *Australopithecines* to the first *Homo* may reflect subcortical rather than neocortical growth. It is this growth that provided the key to hominin survival and, as we will see, also to later growth in the neocortex and the capacity for speech and eventually culture.

Before we can tackle these evolutionary events of dramatically increased emotional capacities, we need to address a prior question: What was available in the phenotypes and underlying genotypes for natural selection to select upon? One way to answer this question is to examine what are termed (1) *preadaptations* and (2) *behavioral capacities*.

Preadaptations

Biologists increasingly use another label, *exaptation*, for the older term *preadaptation*, but I prefer to call them *preadaptations* because it better communicates what is involved. A *preadaptation* or, if you will, *exaptation*, is a trait that evolves as a by-product of selection, producing another trait that enhanced fitness, or the capacity of members of a species to survive and reproduce. Later, after perhaps thousands if not millions of years, this by-product trait, which had simply been sitting there, becomes subject to natural selection because it too enhances fitness of a species under changing ecological conditions. Some of the most important traits of great apes were preadaptations that evolved along with other traits; and then as ecology changed, selection resumed on existing traits or, in some cases, started on a trait that had been simply neutral in terms of fitness consequences. Box 2.1 lists some of the key traits of great apes that also existed in the LCA of great

apes and humans and, thus, that could be subject to selection *if* they enhanced fitness of hominins. I will briefly summarize the most important of
these (see also Turner and Maryanski 2008: 28–59; Turner et al. 2018:
75–105, 128–129).

One preadaptation was a capacity to experience and communicate to
others at least four of the primary emotions: *fear, anger, sadness,* and
happiness. These primary emotions were to be mixed and elaborated upon
to produce more complex and nuanced emotions—much like how the
mixing of primary colors (red, yellow, and blue) can produce many diverse
colors. This process started, I believe, very early in hominin evolution. As I
will outline in the next chapter, this capacity was *the* key preadaptation for
the survival of humans' hominin ancestors, inherited from our common
ape LCA with chimpanzees and other great apes. Social ties are formed
through emotions, as are group solidarities; and thus, for hominins to get
better organized, they had to become more emotional.

Another preadaptation was for *language*, which was critical to the
evolution of the symbolic capacities to produce culture. All great apes

---

BOX 2.1 PREADAPTATIONS AMONG HUMANS' HOMININ ANCESTORS

1. **Comparatively large subcortex and neocortex,** allowing for the enhanced emotions necessary to enhance cognitive capacities and neocortical growth.
2. **Hard-wired capacity for language** comprehension and capacity to communicate at the level of a three-year old human child via the visual sense modality.
3. **Protracted life history characteristics** that involve for long periods of nurturance of infants revealing larger, immature brains at birth.
4. **Mother-infant bonding,** creating potential for formation of nuclear family.
5. **Non-harem pattern of mating,** allowing for choice of mating partners allowing evolution of nuclear family.
6. **High levels of play among young,** thereby increasing capacity to role take and adjust interpersonal responses to conspecifics.
7. **Community orientation** revolving around community a sonly stable unit of social organization, with capacity to mapping of community boundaries and its members, thus allowing for flexibility in meeting selection pressures for new forms of social structure.
8. **Low levels of physical grooming,** thus increasing reliance on interpersonal means of communication through role taking.

today can learn human languages—much as young children learn them—by being immersed in an environment where a language is spoken. Great apes can also learn the sign language used by the deaf, and alternatively, they can learn the meaning of pictograms and then use computer keyboards to assemble these pictograms into messages to their trainers. They can thus "speak" at about the level of a 3-year-old child. Great apes cannot actually speak, of course, because they do not have the physical equipment in key structures—lips, tongue, larynx, and the muscles that regulate them—that allow humans to use phonemes, morphemes, and words strung together by grammars. Moreover, great apes do not have a fully developed *Broca's area*, which is the brain structure that allows humans to download the brain's way of thinking into sequential speech. But *the basic neurological capacity* for language nonetheless existed among the LCAs to great apes and hominins. It was generated as a by-product of natural selection as it began to convert the small, *olfactory*-dominant mammals (as are most mammals) attempting to adapt to the arboreal habitat to *visual* dominance, which would be much more fitness enhancing in a three-dimensional environment where a miscalculation can lead to death by gravity. In so doing, selection created a preadaptation in the neurology of evolving primates for language facility (Geschwind 1965a, 1965b, 1965c; Damasio and Geschwind 1984).

Still another preadaptation is what are termed *life history characteristics*, which are genetically regulated patterns of conception, reproduction, and care during the life cycle. More than any of the other primates, great apes have a longer period of gestation in the womb, prolonged infant nursing by mothers, protracted infancy dependence, and long juvenile phases (Wolpoff 1999; Falk 2000; Dirks and Bowman 2007; Kelley 2004). These life history characteristics were simply part of the nature of great apes, compared to similar-sized monkeys, but they would eventually allow for larger-brained infants to be born early in order to pass through the female cervix. As highly vulnerable and neurologically immature offspring were born, great ape life history characteristics would allow for extended care of neurologically immature infants and children by their mothers and, eventually, by their fathers as the nuclear family evolved. In order for selection to eventually grow the brain, as we know it did, this preadaptation was necessary. Without it, natural selection would have to create behavior propensities for extended infant and juvenile phases of development, which probably would not have occurred if left to natural selection. With the necessary capacities *already present in the genome*, enlarging the brain could occur; and, with a larger brain, spoken language and culture production would begin to evolve.

Another preadaptation is *mother-infant bonding*, which is virtually universal among mammals and was thus available for further selection if the nuclear family and other forms of kinship would be fitness enhancing. Many mammals, and birds as well, evidence a kind of nuclear family of parents and young offspring; and so, it would not be a big stretch for selection to somehow enhance relations among males and females to play parental roles—even if it is not something that great apes normally or habitually do. But, the force behind propensities to create nuclear families was *not* language or large brains but *emotional attachments* that evolved with the expansion of the emotional palette of *Australopithecines* and certainly early species of *Homo* (i.e., *Homo habilis* and *Homo erectus* or *ergaster*).

A related preadaptation is the *non-harem pattern of mating* in which dominant males disproportionately mate with a larger number of females than nondominant males. Monkeys, for example, have a quasi-harem pattern, and in some human cultures harems exist. But, among great apes, males and females engage in promiscuous sex, a pattern of sex that presented problems of how to get males and females committed to each other and to the nuclear family. And, like so much in the evolution of hominins, this potential adaptive problem would be resolved, at least partially, by enhanced emotions rather than by genetically controlled specialized bioprogrammers for nuclear families. Had great apes had a harem pattern of mating, however, the nuclear family would have been much more difficult for natural selection to install. With the nuclear family as *the structural backbone* of the hunting and gathering band and all other forms of human kinship that would evolve in later societies, the lack of genetically regulated harem mating and reproductive patterns among adults opened the opportunity for the evolution of the nuclear family—a structure that is *not* "natural" for great apes and, hence, for humans as well.

Still another preadaptation is for *play*, which is common to mammals but especially important for humans because so much learning about how to interact and to plug into culture depends on infant and juvenile play activities (Burghardt 2005). To have a hardwired propensity to play would allow for learning through practice of key cultural expectations during the prolonged infant and juvenile life cycle phases that are genetically programmed into great apes and, hence, the LCAs to great apes and humans.

Compared to other primates, great apes do not groom as much, with the result that they rely on interpersonal reading of reading of gestures, especially those communicating emotions. As a result, another preadaptation for human interaction and social organization was already wired in the great ape genome: reliance on mutual reading of gestures to form solidarities in group in groups as well as larger social structures.

Another preadaptation is tied to *community organization* (rather than group organization). Animals like monkeys orient only to groups, but great apes orient to a larger social and ecological world of the home range and keep track of who belongs and who does not belong. This orientation, when accompanied by bigger brains, language, and culture, would allow the scale of human societies to grow beyond the here and now of the group. Moreover, because community was the only biologically programmed structural form, beyond mother–young offspring attachments, in the LCAs to great apes and humans, it allowed for groups to form as needed through emotional attachments rather than by genetically installed group bioprogrammers. As a result, hominins and then humans could create diverse patterns of group organization in order to meet changing selection pressures without having to overcome hardwired bioprogrammers for particular types of groups, which is much more typical in the mammalian universe.

The last preadaptation in Box 2.1 is for low levels of grooming and reliance on cognitive mapping of community members. Because of their low levels of sociality, great apes do not groom very much—especially when compared to other primates. Rather, great apes cognitively map the boundaries of their community and remember who belongs to this community. Thus, social relations depend on cognitive skills and use interpersonal skills to form flexible but generally weak-tie relations with community members. Great apes are not genetically locked into particular patterns of grooming, which, in turn, allows them much greater flexibility in forming social relations that would be adaptive to changing habitats.

Behavioral Capacities and Propensities

Great apes and especially chimpanzees, and hence humans' common ancestor with chimpanzees, evidence a large suite of behavioral capacities to act and interact in certain ways as well as a propensity to continuously invoke these capacities in daily life within a community. These capacities rival those among humans. In many ways, they are equal to the interpersonal abilities of humans, even though chimpanzee brains are much smaller than human brains.

At first it might seem counterintuitive to note that animals that reveal predominately weak social ties, that do not form stable groups, and that often walk around alone in their communities would be so facile in such interpersonal behaviors listed in Box 2.2, including picking up interactions with others in a community after not seeing them for some time, engaging in emotion-arousal greeting rituals, engaging in emotional

BOX 2.2 INFERRED BEHAVIORAL PROPENSITIES OF HOMININS

1 **Propensity to cognitively map** the boundaries, membership, and social relations among members within larger, more inclusive communities rather than to form permanent local groupings

2 **Propensity to focus on face and eyes of conspecifics** for assessing emotions during episodes of interaction

3 **Capacity to mimic emotional gestures in face and body** of conspecifics (through activation of mirror neurons)

4 **Capacity to role-take (invoke theory of mind)** to assess the dispositions of conspecifics to act in particular ways

5 **Capacity to achieve emotional empathy** with others during role-taking

6 Propensity to **mimic responses of others** while, at the same time, engage in **role switching, in play activities among the young**

7 Propensity to **fall into rhythmic synchronization of bodies and vocal gestures** during interactions, especially when larger numbers of conspecifics are in propinquity

8 **Propensity for collective emotional arousal** during periodic gatherings of larger numbers of community members and to emit emotionally charged, ritual-like behaviors

9 **Propensity to assess reciprocities in exchanges** of resources with others

10 **Propensity to calculate fairness and justice of exchanges** with others and to sanction (positively or negatively) with emotional intensity those exchanges deemed to be fair or unfair

11 **Capacity to see self as an object in interactions with others** and to emit gestures expressing conceptions of self and to evaluate self by role-taking with others

12 **Capacity to reckon the respective status of self and others** and, thereby, to respond to status differences, particularly those differences marking hierarchy but also those marking distinctive social categories such as age, gender, and community membership

13 **Capacity of males** (only among chimpanzees) **to form friendships** with other males and, occasionally, with favored females as well

Note: By the logic of cladistic analysis, those behavioral propensities and tendencies among great apes provide a good indicator of the behaviors of the hominins, which humans share with great apes. We thus get a glimpse at "human nature" by viewing the behaviors of great apes. This nature is not what is often hypothesized because great apes do not form strong ties, kinship systems, or even permanent groupings; rather they are weak-ties animals oriented to larger communities than local groups. At humans' ape core, then, we are far less social than is normally hypothesized, and this fact gives us purchase in understanding how selection pressures worked to increase sociality and groupness and, in so doing, to create a proto-language of emotions that eventually was blended to a gesture language and that would evolve into an

auditory language as enhanced emotionality allowed the neocortex of late
hominins to evolve to the human measure.

(Adapted from J. H. Turner et al. 2018: 128–29)

rituals of solidarity and celebration when larger numbers of fellow
community members come together, reading each other's eyes (as well as
face and body countenance) for emotional cues, achieving empathy with
others, seeing themselves as objects of evaluation by others in a situation,
reckoning status of others vis-à-vis self, and assessing whether exchanges
of resources with others are fair and just and, in general, being capable of
engaging in highly nuanced and complex interpersonal behaviors.

Yet, with a moment of reflection, it is clear why great apes can
execute these nuanced interpersonal behaviors: the very *lack of bioprogrammers* for forming bonds, kinship, and stable groups. Great apes,
unlike monkeys, do not naturally form strong ties, groups, and kin
units; and thus, they must *actively construct and reconstruct through
interpersonal skills their social world*. Being weak-tied, nongroup, and
non-kin oriented does not mean that chimpanzees are not social or
organized at all. It simply means that they have loose and flexible
patterns of relationships that must be actively worked on rather than
pushed on them by genetically driven biological programmers.

Indeed, unlike mammals with genetically controlled bioprogrammers
guiding the formation of social relations, great ape sociality is far more
difficult to bring off: every interaction is negotiated and revolves around the
mutual exchange of information and emotions where self, others, and situation are salient; where memories of past interactions are invoked; where
the status and number of others co-present are assessed; where the demography of who is co-present is taken into account; and where so many other
contingencies are potentially introduced. If interaction among chimpanzees
or other great apes sound like human interaction, it is; and in fact, what
humans do in interactions is much the same as chimpanzees can do.

Most mammals and birds are driven by "instincts," which are "in their
nature," whereas humans must construct the flow of interaction. They
must also construct the social units—from groups and organizations to
macrostructures—that organize their daily lives. This type of complex and
ever-contingent production and reproduction of our social relations is in
our human (really great ape) nature—dramatically intensified by spoken
language and culture made possible by big brains. But undergirding all
these amazing capacities is an even more fundamental capacity: the ability
to monitor, control, read, understand, emphasize, and otherwise engage in

emotional behaviors because in the end, *it is emotions that make human sociocultural formations either hold together or break apart.*

The list of interpersonal practices outlined will be discussed in detail in Chapter 10 within the *interaction complex*, while the first two preadaptations—emotional and language capacities—will be examined in Chapter 8 within the *emotions complex*. Thus, these two lists of pre-adaptations and behavioral capacities should be kept in mind—indeed, bookmarked in some way—because they will guide our analysis of ape nature as it affected the evolution of human nature.

Conclusion

Thus, we can now begin to appreciate the extent to which human nature is ancient, because it is the inheritance of well over 20 million years of ape evolution. Yet, at the same time, what typifies great ape nature is how loose and flexible it is compared to most other animals. A relatively few powerful bioprogrammers drive ape social relations and social structures. These patterns of weak social ties and loose group formations within a larger community would eventually allow humans to create flexible social structures and patterns of social relations that could be adjusted and adapted to new habitats constantly generating ever-changing selection pressures. It is the human capacity to use a large repertoire of inter-personal skills (inherited from the LCAs of great apes and hominins) to form social bonds, to sustain relations, and to produce and reproduce group formations that makes human societies possible. In other words, human nature is not just a bunch of drives and bioprogrammers, as it is for many animals, including most mammals. Instead, human nature is greatly affected by emotions, language, and large brains able to form culture. This nature is sustained by a set of generalized interpersonal skills and emotional capacities that are actively used to achieve what bioprogrammers or "instincts" do for many other animals.

What makes human nature so complex and difficult to ascertain is the "supercharging" effect of much more complex and intense emotions, bigger brains, spoken language, and culture on aspects of humans' biological nature inherited from the LCAs of the ancestors of great apes and humans. Moreover, the preadaptations and interpersonal propensities of humans that are programmed, to some degree, by our genome can be countervailed, if not subverted, by acts of human agency. Indeed, humans have often created societies that go against what is programmed into the genome inherited from the LCAs of the ancestors to humans and great apes. So, our goal is to unravel as best we can the biology of human nature, even as it is filtered by the biology of emotions, language, and culture. At times it will be difficult to disentangle what is biological and what is socially and culturally constructed, but as long as we remain attuned to what we inherited from the LCAs of great apes and humans, we can overcome the distorting effects of what humans can construct through acts of agency (culture and social structure) on what humans inherited from their LCAs and still carry in their genomes.

Notes

1 Bonobo chimpanzees exhibit a somewhat different community system because of their small habitat, where crowding forces much more contact among conspecifics. The result is tension-reduction rituals, such as the famous genital-to-genital rubbing (GG-rubbing) and considerable sexual contact to create what, on the surface, seem like stronger ties but may only be way for orangutans and humans. This genetic closeness indicates that humans shared ancestors with the ancestor of these presents to manage the stress of being forced together by their restricted habitat. The pattern of common chimpanzees described in the text is, I feel, a better representation of the original chimpanzee social structure, before the subspecies of bonobos broke off from the common line between 2 and 3 mya.

2 A small number of chimpanzees live in a more exposed open-country habitat in west Senegal today, although they must retreat to the trees dotting the savanna-like open-country habitat. These chimpanzees have begun to form more permanent groups for hunting/foraging for food and defense, mostly built around males but occasionally females. Thus, without actual changes in their genomes, it is possible to see natural selection pushing for more group structure in these chimpanzees.

3
Why Humans Became the Most Emotional Animals on Earth

The ultimate ancestors of all humans were rodent-like insectivores that began to ascend into the arboreal habitat after the demise of large dinosaurs about 63–64 million years ago (mya). Adaptation to the arboreal habitat led to the transformation of the sense modalities in these small mammals from olfactory to visual dominance, which in turn rewired the primate brain significantly. And, in the case of the ancestors of present-day great apes, selection had made great apes more intelligent than other primates and had, as a preadaptation, rewired their neurology for the capacity, if ever needed, to use language. Thus, natural selection did not have to create the fundamental capacity for language among hominins; it was already a wired preadaptation as a consequence of the conversion of the brain from olfactory to visual dominance. This conversion involved natural selection moving the dominant sense modality of most mammals, which is in the subcortical areas of the brain (the olfactory bulb), and bestowing dominance to the occipital lobe in the neocortex. In this way, vision became the dominant sense modality of monkeys and apes. Only among great apes, however, did this shift to visual dominance also create the basic neurological capacity for language. The ancestors of great apes today were more intelligent than monkeys, and it appears that this extra intelligence coupled with the rewiring of the areas in and around the *inferior parietal lobe* where the temporal, parietal, and occipital in the neocortex meet, transformed great apes into, potentially, language-using animals.

If we are to understand human nature, then, we first must understand great ape nature, which, as emphasized at the close of the previous chapter, involves analysis of the (1) *preadaptations* in great ape neuroanatomy and (2) *behavioral capacities as action propensities* following from these capacities. By the logic of cladistic analysis, contemporary great apes can provide a look back in time to the characteristics of the last common ancestors (LCAs) of great apes and humans to see what preadaptations and behavioral capacities were available for selection to work on in order to make hominins more fit as they increasingly had to adapt to the terrestrial habitat in Africa and, eventually, a wide variety of habitats in Europe and Asia.

As the cladistic analysis in the last chapter revealed, great apes are fundamentally weak-tie, nongroup, and nonfamilial animals, especially compared to most other mammals and, among primates, compared to monkeys, who are highly organized by female matrilines and male dominance hierarchies in tight-knit group and kin structures. Great apes, then, were just the opposite: organized by only one permanent structure, *the larger community of many square miles*, with fluid, weak-tie relations within this larger community. This mode of adaptation worked well in the marginal niches in the extreme terminal feeding areas high in trees of African forests, where food resources, space, and structural support by branches were scarce compared to the niches occupied by monkeys in core areas of trees in the arboreal habitat. As the forests receded and the secondary forests, bushlands, and grasslands expanded, some arboreal primates had to move into more open-country conditions, where tight-knit social organization would be fitness enhancing.

Monkeys could rather easily make this adjustment because of bio-programmers for female matrilines linking collateral and lineal female kin and dominance hierarchies honed from competition from im-migrating males. In contrast, ancestors of great apes were not well suited to this new habitat and, no doubt, most species went extinct. The ancestors of humans, however, were able to beat the odds because natural selection worked on preadaptations and extant behavioral traits to make humans' ancestor initially more social and, then, increasingly group- and kin-oriented. But how did natural selection get around the liability of weak-tie animals forced onto predator-ridden open-country habitats? It is in answering this question that we get our best look at the underlying biological nature of humans.

Natural Selection and the Forces of Evolution

Natural selection can only work on variations that are present in a species, although the other forces of evolution as conceived by what is called the Modern Synthesis in biology help account for how variations on which selection can go to work are generated. One source of variation is *mutations*, or reshuffling of genes to produce new traits. At one time, early evolutionary biologists in the late nineteenth and early twentieth centuries thought that this force was more important than natural se-lection, but soon it was recognized (e.g., Fisher 1930) and later confirmed by others (e.g., Stebbins 1969) that most mutations are harmful and, thus, do not enhance but decrease fitness, or the capacity to survive and reproduce. Such is the case in complex structures where high levels of interdependence—especially in the mammalian brain—make change generated by mutations in one area harmful to other areas of complex biological systems.

Emphasis of this point highlights a much more important force in the evolution of complex systems: *directional selection* on existing traits that reveal distributions on a bell curve. For example, almost every trait of an animal varies by the fundamental property of size, which will distribute itself in a bell-shaped curve from smaller to larger. If smaller or larger variants on the tails of the bell curves describing their distribution increase fitness, then the favored variant will be passed to the next generation and the less favored variant will begin to disappear. With each successive generation, the favored variant will become more common and, itself, make up yet another, perhaps more tightly structured curve (with smaller *standard deviations* from the *mean*); and, over not too much time, the population will reflect the favored trait across the whole population. Since many of the critical traits—whether preadaptations or behavioral capacities/propensities—in the ancestors of humans were neurological structures and their interconnections as they affected behavior, much of the evolution of humans' ancestors was driven by selection on tails of bell curves describing such traits as size and interconnectivity of neurological structures (Ardesch et al. 2019; Holoway 2015).

Mutations for new traits were thus far less important in understanding the evolution of humans' ancestors. Even fundamental traits that are not so much neurological as anatomical evolved by directional selection more than mutations. For example, as depicted on the cover of this book, great apes can all stand up and, moreover, walk, at least for a time. The anatomical structures involved in this ability—shape of upper legs, ball joints, hips, and other anatomical traits—vary like any other set of traits, with some individuals having an easier time getting and staying upright than others. If being bipedal is fitness enhancing, then selection will favor the ends of bell curves describing the favored tail; and over time, these traits would typify the population, with the less-fitness-enhancing traits disappearing from the genome. Thus, it is *directional selection* on the tail ends of distributions of *existing variants* that can drive evolution, and such was the case as selection began to work on the neurology of great apes to make them more social and group oriented.

Other forces of evolution, such as *gene flow* and *genetic drift*, can also increase variations on which natural selection can go to work. *Genetic drift* occurs because the frequency of *alleles* (variants on genes) can vary randomly. When random variations of alleles (as alleles are separated) take on a "probability value" of being present in the next generation, those that randomly appear and enhance fitness will often be selected, increasing their probability of being passed on and thus increasing their frequency in the gene pool. Over time, the probability of the fitness-enhancing variants being in the gene pool and being selected will increase to the point of significantly altering the gene pool of a population arising out of what began as a random process affecting the frequency of alleles.

Gene flow among animals occurs when individuals of one sub-population of a species move into another subpopulation, with inter-breeding allowing the alleles of one population to flow into the other. Often subpopulations of a species get isolated for a time (as *demes*), and the result is that directional selection and genetic drift change the genome of a population in some way. When members of that population subsequently come into contact with another adapting to a slightly different niche, their genes will mix and change the genome of the two populations (perhaps now joined together as one population).

Thus, the Modern Synthesis sees several sources of variation in a species: (1) current distributions on bell curves of variants, (2) mutations on new alleles or genes, (3) genetic drift, and (4) gene flow. Natural selection "selects" those variants that enhance fitness in a given resource niche. Therefore, when the ecology of a population changes—that is, the environment to which they must adapt to survive and reproduce is altered—then selection may begin to favor certain variants that increase survival and reproduction in the new environment. As the ancestors of today's great apes were forced to adapt to the expanding terrestrial habitats and niches, natural selection began to select on those trait variants of the great ape genome that increased survival and reproduction in this new environment. The preadaptations alluded to in the previous chapter were one source of variation of traits that had long enhanced fitness among primates, and selection began to select variants that would increase fitness in the new, more terrestrial environ-ments. The same would be true for those long-standing behavioral capacities and propensities wired into the great ape genome. Because so many of these capacities and preadaptations involved neurology, much selection was *directional selection*, although some were undoubtedly the result of genetic drift and gene flow and perhaps a few minor mutations.

Natural Selection and Emotions

Expanding the Range of Variations of "Primary Emotions"

Box 2.1 on page 45 provides a list of the basic preadaptations that were subject to selection as the environments of those great apes that would become hominins changed from predominately arboreal to more terrestrial animals. The first and second of these preadaptations are examined in this chapter, with the remaining preadaptations and behavioral capacities/propensities of great apes analyzed in the next.

Virtually all mammals and, hence, great apes have the capacity to experience and express a limited range of primary emotions, such as *satisfaction-happiness, aversion-fear, assertion-anger,* and *disappointment-sadness.* All researchers agree that these four are primary or biology based, but some include additional emotions in the palette of primary emotions, such as *surprise, expectancy, anticipation,* and *disgust.*[1] By primary, it is meant that these emotions are *hardwired into the neuroanatomy* of mammals and great apes. These emotions can be experienced and expressed with varying levels of intensity and nuance. Therefore, the actual number of primary emotions, when variants and intensity are considered, is much larger than just four. For example, *satisfaction* is the low end of intensity of one primary emotion, while *happiness* is the other end of the range of variation. These two emotions express not only different levels of emotional intensity but also somewhat different affect states, as do other valences in between these two extremes. Great apes can feel and project many points in between these poles of a given primary emotion.

We know from many sociological studies that humans create affective bonds, stronger social ties of solidarity in groups, and other types of social structures by the emission of positive emotions such as *satisfaction-happiness* and additional combinations or elaborations of this range (more on combinations shortly). We also know that humans do so through face-to-face interactions, which have the power to charge up the emotional intensity of an interaction (Spencer 1874–96; Durkheim 1912; Goffman 1967; Collins 1975, 2004; Turner 2002, 2007). Thus, it is clear what natural selection did during hominin evolution; selection led to the expansion of the intensity and range of primary emotions into many nuanced variants, thus giving early hominins such as *Australopithecines* (5.0 to 1.4 mya) an increasingly larger palette of emotions to work with, as they sought to form stronger bonds and group solidarities.

This process probably began as soon as *Australopithecines* were fully upright and venturing out onto the savanna for periods of time, beginning as early as 4.8–3.8 mya. At first, time on the savanna was brief, and species of *Australopithecines* moved back to the protection of the forest. But it is clear that *Australopithecines* were the first *hominins,* a term coined to denote any apelike animal near the human line that is bipedal. As emotions began to expand and as attachments and solidarities began to increase among these early hominins, they began to move out for brief times into the open country, away from the protection of the forests.

Being bipedal rather than knuckle-walking was a much more efficient way to get around. Bipedal capabilities were equally or even more important for seeing potential prey and predators since the olfactory sense modality had been reduced when natural selection converted the brain of great apes to be visual rather than olfactory dominant. Moreover, being bipedal frees up the hands and arms for hunting and defense, which would increasingly be the mode of adaptation of hominins as the beginning of hunting and gathering societies emerged later in hominin evolution. This freeing up of the hands and arms from the burden of walking can be seen as another anatomical preadaptation for hunting and defense. Traits honed by natural selection for the arboreal habitat (i.e., brachiation, powerful arms, strong shoulder and wrist joints, and dexterous fingers) could now be used for hunting (throwing spears and eventually shooting arrows) and defense (throwing rocks at predators).

The key to unlocking this potential in ape anatomy, its upper body (i.e., arms, shoulder and wrist joints, and hands with sensitive and dexterous fingers) was tied to the ability to form somewhat stronger social ties and somewhat more stable group assemblages for moving out from the protection of the forests. Alone, a slow great ape is vulnerable to predators and is less able to catch prey—even with a stack of rocks to throw or a lance. In a group, however, *coordinated actions* by great apes could lead to collective food gathering, hunting, and defense by animals capable of throwing spears and rocks, shooting arrows, raising and swinging clubs, and other behaviors that four-legged animals cannot adopt. Thus, bipedalism that allowed for the hands, arms, fingers, and wrists to be free for throwing, grabbing, gathering, and catching actions necessary for securing food and, if needed, for fending off or actually killing predators was accelerated by initial patterns of group organization, where such actions could be now be coordinated and, hence, be fitness enhancing.

It was the initial *expansion of primary emotions* that made these more stable forms of group-level organization possible and, hence, allowed the ancestors of humans to survive (see Table 3.2). This process of forging stronger ties at the group level began long ago, before the hominin brain grew much larger than that of a chimpanzee at about 375–425 cc. Even as brain size moved up to the 500 cc range with early species of *Homo* at around 2.1–2.2 mya (*Homo habilis* and *Homo erectus*), brain growth to the human measure may not have moved much for another million years (see Holoway, 2015, for more details of on brain size of various hominins).

One reason for the modest increase to 500 cc may have been the pushing of the neocortex outward as subcortical areas of the brain expanded first. The data in Table 3.1 compare the size of both neocortical and subcortical

TABLE 3.1 Relative Size of Brain Components of Apes and Humans, Compared to *Tenrecinae*

Brain Component	Apes (Pongids)	Humans (Homo)
Neocortex	61.88	196.41
Diencephalon thalamus hypothalamus	8.57	14.76
Amygdala centromedial basolateral	1.85 1.06 2.45	4.48 2.52 6.02
Septum	2.16	5.48
Hippocampus	2.99	4.87
Transition cortices	2.38	4.43

Sources: Data from Stephan 1983; Stephan and Andy 1969, 1977; Stephan, Baron, and Frahm 1986; and Eccles 1989.
Note: Numbers represent how many times larger than *Tenrecinae* each area of the brain is, with *Tenrecinae* representing a base of 1.

structures in the human and great ape brains, controlling for body size. As is evident, the size of the human neocortex eventually would be three times the size evident in chimpanzees and early hominins. But earlier, millions of years before the neocortex started to grow rapidly, subcortical areas of the brain were growing in order to enhance primary emotions. The measurements of *subcortical* areas of the brain in Table 3.1 indicate that these are twice as large as those among great apes. This difference suggests that natural selection was growing these areas to enhance emotional capacities so that stronger bonds and group solidarities could begin to form among animals without strong bioprogrammers for such relationships. Figure 3.1 highlights the areas of the brain listed under subcortical structures in Table 3.1.

The data in Table 3.1 are old and not collected for my purposes, but they are nonetheless accurate and require some explanation. The numbers in the table represent how many times larger than a simple insectivore, *Tenrecinae*, are various brain structures in great apes (*Pongids*) and humans. This technique is part of the control for body size, but it is also interesting because *Tenrecinae* is probably much like the original mammals that moved into the arboreal habitat after the demise of the large dinosaurs and, thereby, initiated primate evolution. The *diencephalon*, housing the *thalamus*, which routes incoming sensory inputs to both the neocortex and subcortical

Left Hemiphere of the Neocortex:

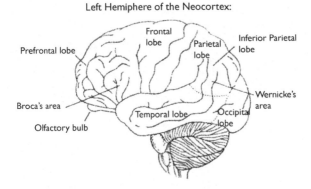

Cross-sectional Analysis of Neocortex
and Subcortical Limbic System

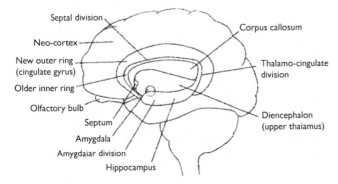

Ancient Subcortical Areas Inherited
From Early Amphibious Vertibrates

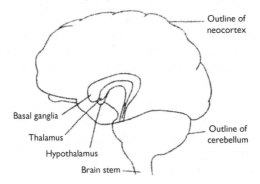

Figure 3.1 Areas of the Brain under Directional Selection

emotion centers, and the *hypothalamus*, which is involved in production of hormones and neuroactive peptides that arouse emotions, are almost twice as large in human than in great ape brains. More revealing is the *amygdala*, which generates *fear* and *anger* in different nuclei; this ancient structure is more than twice as large in humans than in great apes, but most of this difference is in the *basolateral* portion of the amygdala. More recent studies, however, show that it is the *lateral nuclei* more than the *basal nuclei* in the amygdala that are critical because, rather than being related to either *fear* or *anger*, the lateral nuclei are devoted to increasing mutual responsiveness as well as awareness of others and situational cues in face-to-face interactions (Barger et al. 2007, 2012, 2014). Two pathologies to damaged lateral nuclei in the amygdala provide further evidence of what natural selection was doing to hominins. One pathology of damage to the lateral nuclei is autism, where the ability of a person to read the gestures of others for meanings, particularly the emotional meanings and affective states, is limited. The other pathology is Williams syndrome, where the opposite is the case: individuals overempathize with others experiencing negative emotions, so much so that young children with this syndrome will walk up to strangers whom they feel are distressed and hug them.

Clearly, natural selection was adding new nuclei with different functions to the amygdala, perhaps to mitigate the power of raw anger and fear, which disrupts social relations and collective solidarity. The new lateral nuclei, which account for most of the size difference between great ape and human amygdala, work to enhance the capacity of hominins to engage in solidarity-generating interactions by better reading the emotions and intentions of others and the demands of the situational context in which the interaction is occurring.

The more than doubly large *septum* in humans compared to great apes is another case where enhancement of positive emotions revolving around sex has been programmed into humans. The *septum* and other associated areas around it are the source of the pleasure associated with sex. Since great apes are highly promiscuous and enjoy frequent sex, it can be asked, Why would natural selection double the size of the *septum*? It is likely, I think, that the additional nuclei of the enlarged septum in humans represent a way to enhance the emotional experience of sexual partners, perhaps eventually leading to what we know occurs among humans (but not great apes): emotional attachments to sexual partners revolving around emotional variants of happiness, such as *love, caring, compassion,* and *loyalty*. There would be little need to increase the physical pleasure of sex in great apes because they clearly enjoy sex with a much smaller septum and associated neurons. Thus, natural selection

TABLE 3.2 Variants of Primary Emotions

Primary Emotions	Intensity		
	Low	Medium	High
Satisfaction-happiness	Content Sanguine Serenity Gratified	Cheerful Buoyant Friendly Amiable Enjoyment	Love Joy Bliss Rapture Jubilant Gaiety Elation Delight Thrilled Exhilarated
Aversion-fear	Concern Hesitant Reluctance Shyness	Misgivings Trepidation Anxiety Scared Alarmed Unnerved Panic	Terror Horror High anxiety
Assertion-anger	Annoyed Agitated Irritated Vexed Perturbed Nettled Rankled Piqued	Displeased Frustrated Belligerent Contentious Hostility Ire Animosity Offended Consternation	Dislike Loathing Disgust Hate Despise Detest Hatred Seething Wrath Furious Inflamed Incensed Outrage
Disappointment-sadness	Discouraged Downcast Dispirited	Dismayed Disheartened Glum Resigned Gloomy Woeful Pained	Sorrow Heartsick despondent Anguished Crestfallen Dejected

was enhancing the psychological and emotional pleasure of sex, perhaps beginning the process of male-female bonding and, later, commitment between a conjugal couple not only to each other but to the care of their offspring in groupings that began to look like the nuclear family of hunter-gatherers.

The *hippocampus* and attached *transition cortices* are also just under twice as large in humans compared to great apes. These areas of the subcortical portion of the brain are primarily responsible for initial memory formation. The transition cortices hold cognitions about an interaction for the short term during an interaction, and then, if emotions are experienced again, the emotional valancing of cognitions is then stored in the hippocampus (Damasio 1994; LeDoux 1996). If the emotions associated with this cognition are activated in memories, then the cognition is more likely to be remembered for a longer time. The more the same emotions tagging the cognition are aroused and experienced by a person, the more likely is the memory to be shipped to the *frontal cortex* or *lobe* for longer-term storage as a memory (see Figure 3.1). What the functioning of the hippocampus underscores is that memories cannot form unless emotions are attached to them. Research has demonstrated (see Damasio 1994 for illustrations) that when there is damage in the neuro-nets connecting the *prefrontal cortex*, which is the decision-making portion of the neocortex, and *hippocampus*, which is responsible for the initial assembling of memories by attaching emotions to cognitions, a person cannot make "rational decisions"; indeed, this person has trouble making *any* decision. Moreover, damage in the connections between hippocampus and prefrontal cortex affects the ability to remember more generally. Thus, memories cannot form, memories cannot be retrieved, and decisions cannot be made without being tagged with emotions. So, in enhancing the storage capacity of the hippocampus, selection was increasing the cognitive functioning of hominins without initially expanding the neocortex to a significant degree.

Thus, enhancing emotions by initially increasing the range of primary emotions increased intelligence by, in essence, putting a turbocharger (emotions) on the still relatively small neocortex of early hominins. What is also evident is that emotions will have the same effects on all other pre-adaptations listed in Box 2.1 on page 45 as well as on all the interpersonal skills of LCAs of great apes listed in Box 2.2 on page 49. What will become increasingly clear is that the brain cannot grow to the human measure and, hence, language and culture cannot emerge in the human measure *without a prior rewiring of the hominin brain to be more emotional.* For, without increasing social attachments and commitment to groups, which can *only occur via emotions*, the first hominins would not have survived, and thus, it is unlikely that the first *Homo* would have ever appeared.

The emergence and survival of *Homo* required a dramatic expansion of emotional capacities, and the needed emotional turbocharger had to be built into and integrated with the still small neocortex before the larger brain and what it could bring (i.e., full language and robust systems of culture) could be dropped into the cranium of species of *Homo* that would emerge on the hominin line leading to humans. For what is now clear is that intelligence—storage of memories, complex cognitions, and rapid decision making—cannot exist without the evolution of a much larger palette of nuanced emotions. No organic intelligence among any species is possible without a larger palette of diverse emotions; for this reason, intelligent life forms are also highly emotional.

A large neocortex alone would be a protein-draining and calorie-consuming empty warehouse without a larger palette of emotions to tag cognitions so that they can be remembered (stored) and brought to bear (retrieved) in making decisions. Therefore, the neocortex, as reflected in measurements on fossilized crania of hominins, did not emerge until late in hominin evolution. Natural selection was first making hominins more emotional, and in so doing, it was setting the table (as yet another preadaptation) for later growth in the neocortex, for spoken language, and for culture. Humans became, perhaps arguably, the most intelligent animals on earth because our immediate ancestors became the most emotional animals on earth. For all life forms on earth, then, intelligence is a function of emotional capacities, not only in humans but in other intelligent animals such as dolphins, whales, elephants, and some species of birds. Thus, big brains, spoken language, and culture would have to wait for natural selection to rewire subcortical areas of the brain in ways enhancing the emotional capacities of hominins.

How, then, did natural selection enhance emotions? As I have suggested, the first step was expanding the variants of primary emotions, which was done by initial selection on the subcortex where emotions are generated by the release of hormones, peptides, and neurotransmitters. They, in turn, activate the autonomic nervous system, the neurons of the brain, the more general endocrine system, and the musculoskeletal system (see Turner 2000: 84–118, 2007: 43–65 for neurological details). Table 3.2 outlines in a rough form what I think initially occurred, using the minimal set of primary emotions—*satisfaction-happiness, aversion-fear, assertion-anger, and disappointment-sadness*—as a starting point. Initial selection of key subcortical areas worked to simply expand the range of variants of these four primary emotions; and in so doing, it is clear that a much larger palette of emotions can be generated. Moreover, by expanding the one positive among the primary emotions (e.g., *satisfaction-happiness*) natural selection could increase the likelihood that evolving hominins living on the forest floor would form stronger social ties and more stable groupings for eventual brief forays out onto the predator-ridden bushlands and perhaps even the African savanna.

First-Order Elaborations of Primary Emotions

The expansion of primary emotions generates about 75 variants of the primary emotions; and I would hypothesize that initial selection on the emotion centers of the hominin subcortex went this far with *Australopithecines.* This new complex of emotions could allow the neocortex to grow somewhat because of a larger palette of emotions to tag cognitions, and thus, more complex cognitions could be stored as memories and used in decision making. It may be for this reason that during the transition from *Australopithecines* to the first *Homo* (*Homo habilis* and early *Homo erectus*), the brain grew by another 100 cc from 400–425 to 500–525 cc (Haloway 2015). For additional neocortex growth and even stronger social ties and group formations among ever more terrestrial hominins, a next step in the evolution of emotions would require *first-order elaborations of primary emotions.*

First-order elaborations involve a "mixture" of a greater amount of one primary emotion with a lesser amount of another to produce an entirely new set of more nuanced emotions, as is delineated in Table 3.3. Others have made a similar argument (e.g., Kemper 1987; Plutchik 1962, 1980, 2002), but Table 3.3 outlines a set of emotions that has two outcomes: (1) to increase the total number of emotions in the palette, and less successfully, (2) to mute some of the power of negative emotions that dominate mammalian primary emotions (*fear, anger, sadness* generally reduce solidarities, whereas only variants of *satisfaction-happiness* increase solidarity). For mammals with bioprogrammers for group formation, this ratio of negative to positive can be overcome, but for hominins, who are descendants of weak-tie, low-sociality, and non-group-forming great apes, this ratio provides a roadblock to using positive emotions to forge stronger ties in group structures. Such is particularly the case with *fear* and *anger*, which are easily activated because they have a dedicated structure—*the amygdala*—to generate them, whereas sadness is less disruptive because it takes longer to emerge through some combination of neurotransmitter uptake processes and hormones running through the endocrine and neuroactive peptide systems. By my count, around 20–23 of the 50-plus

TABLE 3.3 Combinations of Primary Emotions

Primary Emotions		First-Order Elaborations
SATISFACTION-HAPPINESS		
Satisfaction-happiness + aversion-fear	*Generate*	Wonder, hopeful, relief, gratitude, pride, reverence
Satisfaction-happiness + assertion-anger	*Generate*	Vengeance, appeased, calmed, soothed, relish, triumphant, bemused
Satisfaction-happiness + disappointment-sadness	*Generate*	Nostalgia, yearning, hope
AVERSION-FEAR		
Aversion-fear + satisfaction-happiness	*Generate*	Awe, reverence, veneration
Aversion-fear + assertion-anger	*Generate*	Revulsion, repulsion, antagonism, dislike, envy
Aversion-fear + disappointment-sadness	*Generate*	Dread, wariness
ASSERTION-ANGER		
Assertion-anger + satisfaction-happiness	*Generate*	Condescension, mollified, rudeness, placated, righteousness
Assertion-anger + aversion-fear	*Generate*	Abhorrence, jealousy, suspiciousness
Assertion-anger + disappointment-sadness	*Generate*	Bitterness, depression, betrayed
DISAPPOINTMENT-SADNESS		
Disappointment-sadness + satisfaction-happiness	*Generate*	Acceptance, moroseness, solace, melancholy
Disappointment-sadness + aversion-fear	*Generate*	Regret, forlornness, remorseful, misery
Disappointment-sadness + assertion-anger	*Generate*	Aggrieved, discontent, dissatisfied, unfulfilled boredom, grief, envy, sullenness

emotions in Table 3.3 can be seen as associative, as potentially increasing solidarity and bonds. Others are destructive to social bonds, but many are much like sadness in decreasing desires for strong bonds but not necessarily breaking them (emotions like *dread, weariness, sullenness*). The results, however, are still harmful to social ties, and some are truly negative, as is the case for *vengeance*, which is a mix of happiness and anger. Still, the ratio of disassociative and associative emotions has been reduced, while the nuance of the emotional palette has increased significantly.

The Language of Emotions

The top portion of Figure 3.1 outlines the left hemisphere of the neocortex, with an emphasis on highlighting the principal lobes (*frontal, parietal, occipital,* and *temporal*), the areas around the *inferior parietal lobe* (which give great apes their neurological capacity for language), and the three specific areas responsible for spoken language among humans—Wernicke's area, Broca's area, and the Sylvian fissure. The main lobes and association cortices around the inferior parietal lobe are evident in great apes, as is the Sylvian fissure, but the specific nuclei regulating motor movements along the *Sylvian fissure* for speech and facial movements are not evident among great apes. As emphasized, natural selection prewired great apes for language when it created visual or olfactory dominance. The key to this preadaptation was increasing the number of association cortices in and around the *inferior parietal lobe,* which sits in the posterior part of the brain where the parietal lobe, temporal lobe, and occipital lobe meet. The *olfactory bulb* is subcortical and not under neocortical control and, moreover, the reason that we often immediately experience emotional responses to certain smells. The parietal lobe is devoted to *haptic* (touch, feel, body rhythms) sensory inputs, the temporal is devoted to *auditory* (hearing) sensory inputs, and the *occipital* is devoted to visual (seeing) inputs; and when we touch, hear, or smell something, humans will turn their eyes to the source of the sensory input, seeking a visual image that will dominate over the activation of auditory or haptic inputs. Olfactory inputs are a bit out of the reach of the control powers of the neocortex, but humans still will look in the direction where olfactory inputs are perceived to originate. When olfaction is the dominant sensory input, animals will seek to smell the areas from which the olfactory sensations emanate because that is how their brains are wired.

Sensory input are routed to different nuclei in the *thalamus* and, then, to (1) the relevant lobes in the neocortex for sound (*temporal lobe*), seeing (*occipital lobe*), and feel/touch (*parietal lobe*); and (2) emotion centers in the subcortex, as well as to the *olfactory bulb* also sitting in the subcortex. Because olfactory-dominant animals have their sensory inputs come immediately to the olfactory bulb and the thalamus, the close proximity of the amygdala (fear and anger) allows for a rapid reaction to danger. Moreover, visually dominant humans often react emotionally to visual inputs before they can consciously see them because the thalamus is closer to the emotion centers than the occipital lobe in the back of the brain. For example, a person walking in the forest sensing a gnarly object that signals danger (say, a snake) may find his or her heart already pounding, experience sweating, and be in the process of jumping before this person recognizes the object to only be a fallen branch rather than a snake. Thus, the body systems that generate emotions have already been set into motion before a person is actually "cognitively aware" of what the object is. Such is the speed with which the emotional system in mammals operates, with the full meaning of the object being sensed lagging until the visual input is routed by the thalamus to the occipital lobe for full recognition. Because emotions such as fear are essential to survival (and *defensive anger* if cornered by danger), they can dominate individuals' responses to their environment because they are activated so rapidly and at high levels of intensity. More important to hominin survival than just rapid-fire "fight or flight" responses from danger, however, was the development of stronger social ties as well as commitments to, and solidarity within, groups that, in the words of sociobiologists, could serve as "survival machines" in more open-country habitats (Dawkins 1976).

Thus, emotions will almost always dominate responses that have meaning and significance to individuals, but for low-sociality and non-group-organizing animals, positive emotions were needed to compensate for this lack of strong bioprogrammers for social ties and stable groupings. For some time, I have argued that the prewiring of the great ape brain for language became entangled with "the language of emotions" (Turner 1996a, 1996b, 1997b, 1999, 2000, 2007, 2010, 2020; Turner and Maryanski 2021). As emotions were being enhanced through the creation of more variants of primary emotions and, then, through combinations of emotions during interactions, these emotions were not discrete states as might be implied by assigning them a linguistic label or term listed in Tables 3.2 and 3.3. In reality, emotions flow from eyes, face, and body as *sequences of gestures* that communicate unfolding affecting states. The ordering of sequences of emotional gestures is yet another way of mixing and putting more complex and nuanced emotional states forward to others, thereby creating additional meanings as individuals read each other's gestures during the course of interaction. Moreover, even if the emotions communicated are negative, they can signal internal states of one individual and often encourage sympathetic responses from others. So, negative emotions signaling distress can promote some solidarity with others when the latter are sympathetic, and out of such understanding of another's distress, a more positive emotional bond could be formed. This exchange of more nuanced and complex *sequences of affective states*, both positive and negative, enables individuals to experience intersubjective states as *empathy*—a capacity that great apes already possess (de Waal 1996, 2009, 2019) and thus could be selected (see Chapter 5). Being able to read sequences of affective gestures operates much like "emotional vocabulary" that increases intersubjectivity and solidarity. And, I would argue that this "vocabulary" began to evolve among hominins as they exchanged sequences of nuanced and layered affect during the course of interaction.

I would even argue for a field of "emotional linguistics" in sociology and linguistics whereby the implicit rules of "body language" are discovered. Body language is, of course, not language as contemporary linguists would define language because, first of all, it is more visually based than auditory, although it can contain auditory gestures (such as voice inflections) and, even more significantly, it includes meaningful gestures revolving around touch. Moreover, while it does not have anything that is the exact equivalent of phonemes, morphemes, and syntax ordered by a grammar, it does contain a pattern of ordering of gestures over time in ways that signal conventional meanings about affective states of individuals. Indeed, by reading eyes, face, hands, and body countenance *over time*, emotional meanings unfold and thereby carry common meanings—thus making sequences of these gestures "language-like."

This method of human communication is "body language" today, and many books have been written about how it "speaks" to others and says something about each of us (e.g., Coleman 2005). But more fundamentally for our purposes in understanding human nature, it is still *the primary way in which humans read emotions in, and communicate emotions to, others*, even as talk may appear to be the primary mode of communication. To illustrate the capacity of this language of emotions to communicate common meanings, I often suggest a simple exercise: turn off the sound on a movie and then watch only the cues (facial expressions, body countenance, position of bodies, modes of touch, etc.). Most people follow the general story line of the movie because this was hominins' and humans' most ancient "language," or, let us say, "quasi" or "proto language." And it is *a language that humans still rely on* when trying to pick up the emotional disposition of others. We constantly check verbal utterances about expressed emotional feeling with actual body cues in eyes, face, and body to see if the "utterances" of emotion are, in fact, authentic and genuine. Thus, the primal quasi-language represented by face and body gestures is still critical to forging bonds of solidarity and sustaining the viability of humans assembled in groups. Because overcoming weak social ties and unstable group formations were the driving selection pressures on hominins, and enhancing emotions was clearly the route taken to do so, it seems likely that emotions could be made even more powerful when ordered by an implicit "grammar." Moreover, by being able to read emotions as sequences of body gestures, more varieties of potential emotional states could be built from the primary emotions and their second-order elaborations. Chimpanzees already can read subtle gestures in eyes and face (Menzel 1971), and thus, it is likely that the LCA to the ancestors of chimps and hominins could also do so. The result would be a language-like stringing of emotional signals together in different patterns of presentation to communicate different meanings.

Selecting on emotions would also be relatively straightforward because the subcortical structures of the brain generating them are so ancient and fundamental to survival. If survival depended on a more elaborate system, indeed a quasi-language of emotions, selection on emotions and the in-place neurology for language in the areas around the *inferior parietal lobe* could rapidly increase the intensity, variety, and nuance of emotions needed to build social bonds and create group solidarities.

One of the areas of the human brain important in spoken language is Wernicke's area (see Figure 3.1). It lies on the left side of the normal brain near the inferior parietal lobe and the surrounding association cortices, making language possible. This area is devoted to uploading sensory inputs into the brain for processing; and while great apes do not have an exact replica of this area, there is a homologous area in great apes (Spocter et al. 2010). It is likely that the presence of this homology is what allows great apes to upload languages spoken by humans into great ape brain for processing. In humans, Broca's area is the converse of Wernicke's, and lies in the left frontal lobe (for most people) near the Sylvian fissure. It is devoted to downloading the brain's way of processing information into *sequences of speech production*. Great apes do not possess this area in their brains, and it is one of the reasons that they cannot "speak." They do have what is sometimes termed Broca's hump or Broca's cap, which is an asymmetry on the left side of the brain in approximately the same area as a full Broca area in humans (Falk 2007).

Thus, the basic wiring may have already existed in a rudimentary form as yet another preadaptation related to language and, hence, available for selection if spoken language increased fitness.

The biggest obstacle to the evolution of spoken language was not, however, the basic neurology for language, *per se*, because great apes (and hence hominins) can (could) understand language through the auditory or visual modalities while using their visual and haptic senses to communicate back with hand signals or typed sequences of pictograms to their human handlers. More recent research indicates some pre-adaptations, beyond the Wernicke homology and Broca's hump or cap may have allowed for spoken language to evolve among late hominins. Moreover, the motor control areas along the Sylvian fissure are particularly important in the ability of humans to engage in fine-tuned, articulate speech; and these are not evident in great apes but, as recent data suggest, they may have evolved by the time of *Homo erectus* (1.8 mya).

Additional Preadaptations for Spoken Language

A set of proteins (labeled FOX) constitute a large family of helix transcription factors that target other genes. Of particular significance to the evolution of language are the FOXP1 and FOXP2 genes and their target genes as they affect humans' capacity for finely articulated speech (Crespi, Read, and Hurd 2017; Schulze, Vargha-Kadem, and Mishkin 2018; Wohlgemthuth, Adam, and Scharff 2014). It is clear that these genes have undergone evolution during hominin and human evolution. In particular, FOXP2 proteins are involved in developing and coordinating the vocal tract and the muscles affecting the mouth and lips for speech; damage to these genes causes speech impairments. At one time, it was thought that selection had been working on these areas for only 200,000 years (Enard et al. 2002a, 2002b), but it now appears that FOXP2 existed among *Homo erectus* as early as 1.8 mya. Thus, it is reasonable to view FOXP2 as a preadaptation for spoken language (Enard 2007, 2016). It is not known whether humans' most immediate ancestor possessed the capacity for finely articulated speech in the human measure, but even if early *Homo erectus* did not, the existence of FOXP2 (and FOXP1) could be subject to selection, if spoken language had fitness-enhancing consequences. Great apes and hence humans' hominin ancestors possess(ed) the capacity for calls and vocal sounds, some of which can be unique to particular populations of apes. Yet, apes are limited by the lack of an open vocal track and the muscles in and around the mouth, larynx, and lips that allow for rapid, fine-grained speech articulations.

As noted, great apes possess a "hump" or "cap" in the approximate location of Broca's area in the human brain, and so, like FOXP2, this hump could be subject to directional selection if downloading thoughts into speech was fitness enhancing. The existence of Broca's hump and FOXP2 means that radical and potentially harmful mutations (Fisher 1930; Stebbins 1969) would not be necessary if selection was pushing for speech. Rather, directional selection on the fitness-enhancing tails of the distribution of FOXP1 and FOXP2 alleles as well as the distribution of those hominins with a larger Broca's hump could, over generations, lead to greater capacity for speech in the human measure. Because the basic neurological wiring for language in general (as is evident among the great apes today) was already present in the neurology of great apes millions of years ago, something as complex as speech could evolve without the need for harmful mutations simply by directional selection on the favored tail of bell shape curves. Clearly, such has been the case: humans can talk, and they can talk to great apes and receive a reply via great apes' visual and haptic sense modalities (e.g., sign language or typed pictograms).

How, then, did the syntactical and grammatical aspects of speech in the human measure evolve? The first language among hominins was gestural since present-day great apes clearly employ not only calls (Bornkessel-Schlesewsky et al. 2014) but a relatively complex system of hand and body gestures that carry common meanings (MacSweeney et al. 2008a, 2008b; Corballis 2009, 2017a, 2017b; Smith and Delgado 2015; Call and Tomasello 2008). The fact that great apes have little trouble learning American Sign Language reinforces the conclusion that great apes already have the neurological wiring for language. They can upload signs as well as other sensory inputs via the homologue to humans' Wernicke's area into the brain for processing, and they can download their thoughts via hand signals to conspecifics, and at times, simple vocal calls. Without an enlarged Broca's area and FOXP1 and FOXP2 of humans, they cannot engage in the rapid-fire, fine-tuned speech production evident among humans. Because Broca's hump in chimpanzees (and, no doubt, in humans' shared hominin ancestor) is (was) already activated during communication (Taglialatela et al. 2008), directional selection could work rapidly to grow the hump into Broca's area, if speech production had fitness-enhancing value.

This initial pro-gestural language was critical to ancestors of con-temporary great apes and, hence, was essential to early hominin commu-nication. But by itself it could not, I believe, lead to high levels of solidarity and the more permanent group formations that were critical to hominin survival in predator-ridden, open-country habitats. As already emphasized, selection increased the size and connectivity among subcortical areas of the hominin brain by enlarging subcortical centers where emotions are generated, beginning about 5 mya, if not before. At some point in this evolution of increased emotional capacities, the power of the existing gestural signals (perhaps even a proto-gestural language) was dramatically enhanced by their association with emotional gestures, signaled by the face and eyes, in conjunction with body movements and positioning. Nonverbal gestures could now carry more complex meanings and, most importantly, meanings that encouraged more intense emotional bonds that would lead to better organized and more permanent group formations and eventually evolve into nuclear families in hunting and gathering bands. Calls and other vocalizations would also be enhanced by their association with emotional gestures, thereby increasing collective solidarities even more.

Selection, as it pushed for enhanced emotionality attached to gestural signals and auditory calls, increased fitness as hominins began to adapt to bushland and other more open-country, nonarboreal habitations. Eventually, hominins could venture out of Africa to a range of new habitats in Europe and Asia. As connecting emotions to hand and body gestures as well as to vocalizations proved fitness enhancing, selection would continue to expand the emotional repertoire and thereby give almost all signs emotional valences along with whatever other meanings these gestures and calls were intended to communicate. Thus, while we often think of "body language" as an adjunct to spoken language, an evolutionary perspective can reverse this thinking. The underlying neurology for the "language of emotions" *was already in place* among the ancestors of great apes and hominins, allowing for the ordering gestures carrying emotions to be signaled to others. Without this prior emergence of an emotional linguistics to forge stronger bonds and solidarities, spoken language would never have evolved because hominins would probably not have survived in the more open-country habitats to which they were *forced* to adapt. Thus, once fitness was increased by the solidarity-generating effects of emotions, fitness could be increased even further by a spoken language and the use of speech to build up culture. Moreover, the brain could now begin to grow because emotional tags could be attached to all types of cognitions and speech acts. This codification of speech into cultural codes carried affective meanings.

The brain had already grown to at least 500 cc with early *Homo* some 2.5 to 2.2 mya, and perhaps it continued to grow up into the lower ranges of the human neocortex as selection favored a larger brain and emotion-laden gestures and calls. My view, however, is that it was perhaps only over the past 500,000–700,000 years that emotionally charged gestures and calls had reached the point where the brain could begin to grow from the 500 cc to 1,050 cc range, but whether such is the case is less important than the coupling of gestures, calls, and emotions and, thereby, their expansion into a larger repertoire of signals with linguistic features of true languages (such as American Sign Language). Indeed, American Sign Language is a fully developed language emitted by signs in proximity to face and eyes communicating emotions, and thus, operates much like the spoken language of humans where inflections of voice, along with gesturing from facial, eye, and body movements and positioning in interpersonal encounters communicate complex, emotionally charged meanings.

The defining characteristic of hominins is their bipedalism, which can be seen as yet another preadaptation for an enhanced gestural language, because bipedalism freed up the hands for communication and allowed individuals to stand or sit *face to face* and to communicate more complex and nuanced emotions read though the visual sense modality. As gestures became organized into a true language capable of carrying complex meanings and emotions, selection began to push for not only a larger brain that would enhance the language of emotions and gestures, but moreover, selection began to work on Broca's area and FOXP2, in particular (if necessary), to enhance the ability to download thinking into finely articulated speech capable of communicating both instrumentality and sentimentality via voice inflections, coupled with the face, eyes, and body gestures in order to reveal emotions in their most robust form. In turn, this enhanced capacity to communicate emotions would push selection on emotion centers further; and in turn, this enhanced emotionality would make growth of the neocortex ever more fitness enhancing.

Just when and where a more sophisticated gestural-visual language facility first emerged is unknown. It could have been relatively early in the evolution of hominins (6.0 to 4.5 mya), or it could have been relatively later in hominin evolution (4.0 to 2.0 mya) with the emergence of *Homo erectus*. In any case, it is likely that late *Homo erectus* (0.6 mya) revealed a human like language, using the expanded neurology of a brain at the lower range in size (1,050 cc) and intelligence of the human brain (1,350 cc). Thus, humans and their often-heralded capacities for abstract thought, speech, and culture are not uniquely human but, rather, extensions of an evolutionary trend that began with early *Homo erectus* and then accelerated in the second half of *Homo erectus*'s time on the planet right up to the emergence of early *Homo sapiens* 450,000–350,000 years ago. But even this late arrival of language in the human measure was built on the platform of the language of emotions, which probably began at least 5.0–4.5 mya.

Brain Growth, Second-Order Elaborations of Emotions, Spoken
Language, and Culture

Even though early hominins possessed the neurological wiring for language
in general (as do all great apes today), an incipient Broca's "hump" a
homology to Wernicke's area, and a system of calls and gestural communication, the evolution of human language would *not* have occurred without
selection finding a way *to increase sociality among weak-tie primates that did*
not form high-solidarity and permanent groupings. All hominins could have
gone extinct without selection hitting upon a solution—the enhancement of
emotions—among the rather low-sociality hominins' capacity to form more
cohesive and stable group formations. Once selection began to grow the
subcortical areas of the brain, the language capacities of hominins could be
unleashed as selection could successively or simultaneously link emotions to
an expanded gestural system of communication, grow Broca's hump, en-
large the neocortex, push FOXP2 to allow for articulated speech, and
thereby make late hominins capable of spoken language. Thus, growth of
the subcortical emotion centers of hominins' brain was *the driving force* in a
chain of events that allowed hominins to get better organized and, then,
increase their capacities for language to the point where growth in the brain,
spoken language, and culture could emerge.

While it may seem, at first, that expanding the emotional capacities of
hominins could not possibly be the key to the evolution of linguistic
great apes using culture, it is important to remember that language and
culture have to be built upon a neurological platform that would make
growth of the neocortex fitness enhancing. Once this simple criterion is
invoked, emotions become the key breakthrough, but only if the neurological capacity for language already existed in the 400 cc brains of
early hominins. Part of human nature, then, revolves around what the
larger brain does and how it affects other preadaptations and behavioral
propensities—as we will see in later chapters. The evolution of the brain
from 400 cc to an average of 1,350 cc was only possible if a larger range
of nuanced and complex emotions had *already evolved* and allowed for a
larger and more complex range of cognitions to be remembered and to
be used in decision making. Moreover, because the expansion of emotions was, I believe, piggy-backed onto the existing neurology for language still evident among great apes and, hence, the LCAs of great apes
and hominins, the language of emotions became *the* platform for
building up gestural languages and spoken languages. Without this prior
neurological wiring for language as it affected the way in which emotions
evolved as sequences of affective states communicating common

meanings, spoken language could not have evolved. Big brains do not necessarily guarantee language, although most intelligent animals on earth clearly have some system for complex communications. A spoken language, however, has special characteristics. Once the uttered symbols become arbitrary signs of common meaning, communication is dramatically enhanced and sets the stage for the production of culture (as symbolically articulated common meanings can become quite complex). Human language allows for communication at many different levels, from the immediacy of a present situation to highly abstract conceptualizations, whether these be science, religion, or some other accumulated body of information guiding human conduct. Culture in the human measure could not emerge without a language revealing these qualities; and without culture, symbolic forms of social control could not evolve. Norms, values, beliefs, ideologies, and the many other symbol systems used by humans to regulate their social relations would not evolve; and moreover, they would not have any "teeth" or capacity for social control without the emotions attached to cultural codes. *Emotions are what give cultural instructions their power*, with positive emotions and sanctions arising from a person and others because of conformity to cultural dictates and with negative emotions and negative sanctions becoming evident when dictates are ignored or violated.

The key emotions giving culture this power to control and regulate appear to be unique to humans, at least compared to great apes. There is, I argue, a *second-level elaboration* revolving primarily around the emotions of *guilt* and *shame* delineated in Table 3.4 (Turner 2002, 2007; see also Boehm 2013). *Guilt* is the emotion experienced when a person has violated a cultural and moral instruction, whereas *shame* is the emotion felt when a person senses that they have behaved incompetently in the

TABLE 3.4 The Structure of Shame and Guilt

Emotion	Rank-Ordering of Constituent Primary Emotions		
	1	2	3
Shame	Disappointment-sadness (at self)	Assertion-anger (at self)	Aversion-fear (at consequences for self)
Guilt	Disappointment-sadness (at self)	Aversion-fear (at consequence for self)	Assertion-anger (at self)

eyes of others (in relation to expectations of others and expectations contained in cultural codes). These emotions of social control are what allow societies to be built up in the human measure.[2]

In Table 3.4 the properties of *guilt* and *shame* are described. Each of these critical emotions is built, I hypothesize, from the three negative primary emotions: *anger, fear,* and *sadness.* It is the ordering of these three emotions that, I believe, distinguishes the experience of *shame* from *guilt.* In both emotions, *sadness* is the dominant emotion; and so, it is the order of magnitude of *anger* and *fear* that makes the critical difference. In *shame, anger* at self for being incompetent is the second-most powerful emotion mixed with sadness, whereas in guilt, *fear* about the consequences to self of violating moral codes is the second-most powerful emotion after sadness. And so, for *shame* the third emotion is *fear* of the consequences to self for behaving incompetently in the eyes of others, whereas for *guilt* it is *anger* at self for violating moral codes. These *second-order elaborations of emotions* are only possible with culture because, in both, cultural expectations are articulated with spoken language—whether as manifested in the expectations of others or as enshrined in moral codes (norms, values, beliefs, ideologies, etc.); they are the referent for these emotions of social control. *Shame* and *guilt* could not, I believe, have evolved without the elaboration of primary emotions and, then, the creation of second-order elaborations as a neurological base. As a consequence, the brain was allowed to grow with these earlier emotional elaborations and natural selection hit upon (by chance) yet a further solution to social control: combining the three negative emotions to produce great psychological pain for individuals in the name of increasing conformity to group and cultural expectations that were evolving as language was moving from the language of emotions to gestural languages and, then, to spoken language.

Once *guilt* and *shame* are in place, they encourage the further elaboration of cultural codes because they can now be "enforced" by powerful negative emotions and subsequent pain that individuals wish to avoid. External social control involving negative sanctions by others, which can often arouse *counter-anger* and thereby disrupt social relations, is increasingly complemented by self-sanctioning by individuals as they experience *shame* and *guilt.* It becomes possible to have social control without so much *counter-anger* arising from external negative sanctioning by others, thereby enabling groups to sustain their solidarity even as individuals engage in internal *self-control.*

Yet, somewhat ironically, as *shame* and *guilt* become powerful emotions of social control as part of human nature, the growth of the brain and rewiring of subcortical areas and their relation with the prefrontal cortex as well as the frontal lobe, where longer-term memories are stored, allow for the activation of what are viewed as *defense mechanisms* by psychoanalytically oriented scholars. Defense mechanisms allow individuals to push below the level of unconsciousness painful emotions of any kind but particularly *guilt* and *shame*. Emotions that attack self are painful. The result of the brain's effort to avoid this emotional pain is to disrupt the links between (1) the *prefrontal cortex*, on the one side, and (2) the *hippocampus* as well as (3) the *frontal lobe*, on the other side. Just as the body will go into shock with extreme physical pain, so the brain will often block the full impact of guilt, shame, and other negative feelings about self by disrupting connections between decision-making consciousness (by the prefrontal cortex) and memory formation (by the hippocampus and frontal lobe). Table 3.5 outlines some of the most prominent defense mechanisms.

I view *repression* as the master defense mechanism because it removes full awareness of the negative emotions and feelings about self, at least to a degree. The other defense mechanisms generally generate one of the negative emotions or, as is the case with sublimation, produce positive emotions that are experienced by a repressed person. *Anger* is the most likely emotion to be experienced, but *fear/anxiety* and *sadness/depression* can also emerge. The key is that the full impacts of *shame* and, in particular, *guilt* are mitigated. Thus, a large brain can also evolve to reduce the power of key emotions essential for social control that "protect" the person but may still disrupt social relations if *anger, fear,* and *sadness* become chronic as the only emotions experienced by a person and released during interaction with others. In the simple hunting and gathering societies of early humans, repression was less needed because of the simplicity of social relations, but as societies have grown in complexity and, as a result, have elaborated cultural codes codified into moralities, expectations on individuals have increased and, when violated, will be negatively sanctioned by others or explicit agents of social control and, as a result, lead to the activation of one or more of the defense mechanisms delineated in Table 3.5. Thus, activation of defense mechanisms becomes a part of humans' evolved nature as a consequence of emotional enhancements and elaborations, brain growth, and the emergence of spoken languages that can specify and delineate systems of moral coding in culture. Thus, extreme emotionality, articulation of moral codes, and activation of defense mechanisms become a part of human nature, built upon the base of emotions and wiring for language evident in the LCAs of present-day great

TABLE 3.5 Repression, Defense, Transmutation, and Targeting of Emotions

Repressed Emotions	Defense Mechanism	Transmutation to	Target of
Anger, sadness, fear shame, guilt, alienation	Displacement	Anger	Others, corporate units*, and categoric units**
Anger, sadness, fear, shame, guilt, alienation	Projection	Little, but some anger	Imputation of anger, sadness, fear, shame or guilt to dispositional states of others
Anger, sadness, fear, shame, guilt, alienation	Reaction formation	Positive emotions	Others, corporate units, categoric units
Anger, sadness, fear shame, guilt, alienation	Sublimation	Positive emotions	Tasks in corporate units
Anger, sadness, fear shame, guilt, alienation	Attribution	Anger	Others, corporate units, or categoric units

Source: J. H. Turner, *Human Emotions: A Sociological Theory* (2007).
Notes:
* Corporate units are structures revealing a division of labor geared toward achieving goals.
** Categoric units are social categories that are differentially evaluated and to which differential responses are given. Members of categoric units often hold a social identity.

apes, hominins, and humans. They are part of human nature because they are biologically wired into the neurology of humans.

Conclusion

The two preadaptations examined in this chapter—neurological wiring for (1) emotions and (2) language—allowed new elements of humans' biological nature to evolve. Indeed, once emotions increased fitness (by increasing social bonds and group solidarities) among hominins in more terrestrial and open-country habitats, a critical chain of events was set into motion. The direction of natural selection would lead to not only the elaboration of emotions but to brain growth and language, which, in turn, would make social control increasingly cultural and emotional among evolving hominins and, thereby, set the stage for further elaborations of human nature.

In the next chapter, I examine what was perhaps the most startling invention of late hominins or, perhaps, only early humans: *the nuclear family*. As emotions became the basis for communication, social control, and increased solidarity and bonding among low-sociality hominins, fitness was enhanced among hominins living and seeking to adapt to more open-country habitats. Grouping began to emerge, but the most important grouping, which is natural for a great ape, is the creation of a stable reproductive unit consisting of mother, father, and offspring. If selection had not pushed hominins to form these reproductive units, it is doubtful that they would have survived, much less migrated out of Africa to Asia and Europe. Thus, to understand human nature, we need to understand how something *not in the nature* of LCAs to great apes and hominins could evolve and become the structural base of the first societies among late hominins and early humans.

Notes

1 See Turner (2000: 15–16, 2007: 4–5) as well as Turner and Stets (2005: 12–13) for an inventory of key scholars' various lists of primary emotions.

2 De Waal (2019: 121–171) suggests that chimpanzees may experience not only what I have termed many *first-order elaborations and combinations* (Table 3.3), but, moreover, even these second-order emotions in rudimentary form. Thus, it is possible that natural selection did not have to work especially hard to install these in humans.

4
Why and How Did the Human Family Evolve?

Among most mammals, the organization of each species revolves around solving the problem of reproduction. As Table 2.1 on page 35 outlines in brief, great apes and, hence, the last common ancestor (LCA) of great apes and humans had little structural basis for reproduction. Adult males and females were promiscuous; and except for a few cases, adult males and adult females did not form permanent social relations. Thus, paternity was never known in humans' distant ancestors, and biological fathers were not involved with the care and raising of their offspring. Only mothers and their preadolescent offspring would form a strong bond, which was broken at puberty as both sons and daughters transferred away from their mothers for the rest of their lives—at least before the ancestors of gorillas and chimpanzees began to settle in a more terrestrial habitat. Yet, the nuclear family, which early sociologists like Auguste Comte (1830–1842) saw as the basic building block of society—indeed the functional equivalent of the "cell" in the societal "organism"—stands in stark contrast to the reality of humans' inherited biology.

How, then, did the nuclear family ever evolve when paternity was not known and offspring left their mothers at puberty? The result of these biology-driven activities was to cut off lineal ties across generations and even ties of offspring to their parents and to their siblings. The nuclear family is, therefore, *not* natural to humans as many would think, but rather had to be constructed for hominins to survive and take up hunting and gathering without powerful bioprogrammers pushing for this pattern of social organization.

The only relatively stable social structure organizing the LCAs of humans and great apes was the larger community or home range that could be many square miles. This structure appears, if chimpanzee behaviors are any guide, to have been defended against incursion by other males. Community members knew the boundaries of their community and the demography of who should be present. Within the community, however, there were no permanent groups among the LCAs. Although periodic parties could hook up, only to disperse, many individuals wandered around their community alone or occasionally with another for a brief time.

Mothers and offspring were, therefore, left to fend for themselves within the community, with fathers never being known and other adult females as relative strangers to each other because of their immigrant status as refugees from other communities who, of course, were welcomed by males to replace the females who had been born in a community and left at puberty. New immigrant females were simply tolerated by other females, and females would sit together to let their children play. Yet, the community's females did not generally form close relations because they had not grown up together and were, in essence, strangers to each other.

What is even more remarkable is that as the brain began to grow with later species of hominins, infants had to be born earlier and were, therefore, less neurologically developed, making them highly vulnerable. If one were living during the time of the first hominins, it would be easy to predict that they would not survive open-country terrestrial habitats, and yet they did. Humans' nature is thus tied to *how hominins were able to create the nuclear family and more stable group structures* within the larger community or home range.

This pattern of weak social ties and lack of permanent group-level structures, especially kinship structures, is rare among mammals; and as we will see in the next chapters, it helps explain the behavioral capacities and propensities of the LCAs and humans today. In this chapter, our goal is to lay out in more detail the preadaptations for the kind of kinship systems that did eventually emerge among late hominins, and, more generally, the nature of group formations. Because emotions were the bonding force driving these formations, it should not be surprising that human families and group structures in general are often unstable and blown apart by negative emotions. Rather than see these as pathologies, we should recognize that such instability is *part of human nature*, inherited from low-sociality and non-group-forming LCAs that were forced to become group oriented or die.

Community as the Structural Basis of Social Organization

Community as the Natural Social Form

Humans have a natural tendency to have a sense of community. We often think fondly of our "hometown." We take "pride in our community." And, there are occasional fads where we inform others of "our community" (like the oval stickers that were once very popular with abbreviated letters symbolizing "our city" of residence, much like those for countries on European cars). In fact, this propensity to reckon community seems natural because it is in our nature, being inherited from LCAs that go back at least 16 million years. Identifying with community is often more automatic than with groups and organizations because stable groups and organizations are, in evolutionary terms, recent constructions that are not driven by bioprogrammers the way orientation to community is.

While this orientation to community worked against stable kin units and groups among the ancestors of great apes and hominins, it would become a preadaptation for what human societies would become: huge, populated by millions of individuals. No other mammal can come close to this kind of scale in societal organization, primarily because they have powerful bioprogrammers for kin and group organization that limit their horizons. For a large mammal like humans to be able to construct societies on the scale of tiny insects is a rather remarkable achievement, even if it poses danger to the world's ecosystem. With 10 billion humans on earth soon to be a reality, our extinction and the extinction of many other species may be inevitable. Still, we should ask, How have humans been able to build such large societies? It begins with a community rather than with kin or group organization. And, as we will see in later chapters, the mega societies that humans now live in may be more "natural" to an evolved ape than those societal formations—horticultural, pastoral, and agrarian societies—that evolved after hunting and gathering was displaced by more settled forms of social organization.

Low Levels of Grooming and Reliance on Cognitive Mapping

Among many species of primates, grooming is a mechanism for increasing and sustaining the strength of social ties and group solidarities. Among great apes, however, grooming is rather minimal because of the weak-tie, nongroup nature of great ape communities. Robert Dunbar (1996 [1984]) has argued that language replaced grooming when groups became larger, but this line of argument ignores two facts. First, great apes and hence hominins do (did) not groom much compared to monkeys, who are more group oriented. Second, great apes are *not* organized into groups, as are most mammals, but communities spread out over several and, often, many square miles. Moreover, Dunbar assumed that 100–150 individuals is about as many as one can have in a grooming population, and while this may or may not be true, great apes have little trouble remembering this many individuals in their community. When great apes meet up, they engage in interaction rituals signifying mutual recognition and, it appears, interact rather easily, seeming to remember the last time that they interacted. Thus, even with their relatively small brains (375–400 cc) great apes can cognitively map their community, both its geographical boundaries and its demography. This propensity to not only reckon community but to conceptualize this community in more abstract terms is, I think, yet another preadaptation for humans' eventual capacities to hold conceptions of larger symbolic communities in their minds and to respond to expectations of these larger, more remote communities. Humans can identify with not only ethnic communities but also nations as a whole. This capacity to conceptualize more remote spaces and populations as part of "one's community" would be an important capacity when societies began to grow. Even among simple hunting and gathering bands of several nuclear families, the larger system of bands speaking a common language could be conceptualized even if actual interactions with these remote "others" were rare.

It is the ability, then, to look beyond the local group that is critical, not only for being a member of a larger society but for being a member of a larger category of others, such as fellow ethnics or fellow religious worshipers. Indeed, conceptualizing a supernatural realm inhabited by sacred forces and beings is but an extension of the basic tendency among small-brained great apes to see beyond the present and the immediate (Turner et al. 2018). With a larger brain, the horizons of late hominins and early humans would be dramatically expanded.

Yet, the immediate problem facing hominins was not to organize mega societies but to form kin units able to protect young and vulnerable offspring, to form groups to coordinate food gathering and hunting, and to protect group members from predators. An orientation and cognitive mapping of community would prove necessary when larger-scale societies began to evolve; these were not the cognitive capacities that hominins needed as they sought to adapt to more terrestrial environments millions of years ago. This orientation to community, however, did not restrict the nature of the kinship system nor the forms of group structures that would evolve. In a sense, the lack of bioprogrammers for *particular types* of kin relations and groups allowed natural selection room to pursue a variety of strategies before hitting upon the one that would enhance the fitness of early hominins. In the last chapter, we saw the route that natural selection began to pursue: increasing the range, varieties, and nuance of emotions to forge stronger bonds that, in turn, would eventually allow the brain to grow and thereby make spoken language and culture fitness enhancing. Thus, emotions were the key to filling in needed kin and group structures among community-oriented hominins, if hominins were to survive. We look at the other preadaptations (listed in Box 2.1 on page 45) in this context, where natural selection hit upon the solution to the problem of making a weak-tie, non-group-forming animal more social and group oriented.

Life History Characteristics of Great Apes

If emotions, bigger brains, spoken language, and culture were the route that natural selection would take to make hominins more social and group oriented, then dramatic changes were necessary in either the anatomy of female hominins, if infants with large brains were to pass through the cervix at birth, or in some other organizational pattern that could protect infants born with undeveloped brains capable of fitting through the cervix before reaching their full size. Fortunately, selection did not need to work much on the anatomical limitations of the female cervix because great apes already had life history characteristics that would lead mothers to care for their offspring with neurologically immature infant brains. *Life history characteristics* are those related to developmental patterns of a species, particularly length of gestation in the mother's womb, age at weaning, speed of maturation, age of sexual maturity, rate of reproduction, interval between offspring, and length of life cycle (Kelley and Smith 2003). This approach can be used to study living species, but it is also useful when examining fossil remains of species that are now extinct (Nargolwalla et al. 2005; Kelley 2004).

All primates mature slowly and live longer than most other mammals, and great apes do so more than all other primates. Table 4.1 compares adult male baboons (large monkeys) and adult male chimpanzees in terms of key life history characteristics (Wolpoff 1999; Falk 2000). Chimpanzees are much bigger than baboons, which is the case with all great apes compared to the largest species of monkeys, but it is the life history characteristics that are quite revealing.

Gestation for chimpanzees and all great apes is 53 days longer. Nursing for great apes is 1,040 days, which is 1.5 years longer than the baboon. The juvenile phase is 2.6 years longer. The adulthood phase is a decade longer for chimpanzees compared to baboons (34 vs. 23 years), and the spacing of births is almost four years longer. One might think that the greater size of chimpanzees may account for these differences, but these life history characteristics *are clearly built into the ape genome,* as is evident if we use small gibbons (apes, but not great apes in either size or intelligence). The Asian apes, various species of gibbons, are about 15 pounds as adults, and yet they too reveal prolonged life history characteristics: 205 days in gestation, 730 days of nursing, two years in infancy, 6.5 years in the juvenile

TABLE 4.1 Life History Characteristics of Monkeys and Great Apes

	Male Baboon (Monkey)	Male Chimpanzee (Great Ape)
Average weight	50 lbs	115 lbs
Gestation	175 days	226 days
Nursing	420 days	1,460 days
Infancy	1.6 years	3 years
Juvenile phase	4.4 years	7 years
Adult phase	23 years	34 years
Spacing of births	1.7 years	5.6 years

Sources: Wolpoff 1999; Falk 2000.

phase, spacing of births is 2.7 years, and adulthood of 23 years. Thus, size alone cannot account for the differences between chimpanzees and baboons when the comparatively small gibbon also evidences longer life history characteristics. A genetic-based set of prolonged life history traits is evidenced in the hominoid (ape and human) line.

Fossil records confirm this conclusion. For example, examining the teeth of fossils at various ages at death, a great deal can be learned about life history characteristics. Teeth are often well preserved, and their growth is under genetic control and hence can serve as reliable markers of life history characteristics (Dirks and Bowman 2007). For example, by examining the teeth of *Victoriapithecus* (who lived 19 million years ago [mya]) with those of living old-world monkeys today (Dean and Leakey 2004), dental development remained the same, which suggests that life history characteristics of monkeys are under genetic control. In contrast, if we examine several Miocene hominoids, including *Afropethecus turkanensis* (17.5 mya) with *Sivapithecus parvada* (10 mya), the eruption of their molars falls far outside of the range for old-world monkeys. Indeed, the Miocene hominoids have about the same pattern of molar growth as contemporary hominoids such as chimpanzees. Thus, for well over 20 million years, apes, hominins, and then humans have had these extended life history characteristics, which probably evolved in the arboreal habitat when apes dominated. Having plenty of food and protection from predators presumably eliminated selection pressures to speed up reproduction. This would change, of course, as monkeys began to take over the verdant core of the arboreal habitat, thereby selecting out apes and pushing those that survived to the terminal feeding areas of the arboreal habitat.

Once natural selection began to expand emotions of hominins to forge stronger social ties, kin groups, and stable groups in general, selection along these lines would eventually allow for the growth of neocortical portions of the brain. Enhanced emotionality allowed for cognitions to be tagged with a larger palette of emotions, thereby making hominins much more intelligent. The brain did not grow significantly, however, for at least 2.5 million years of hominin evolution, but when it did, a major obstacle to such growth was mitigated by the life history characteristics of great apes who were morphing into hominins. Infants could be born with incomplete growth in their brain that allowed them to pass through the cervix. The long nursing phase, even longer infancy phase, and prolonged juvenile phase, coupled with a spacing of births at five to six years, helped ensure that mothers could adequately care for their offspring. Natural selection could not have installed these life history characteristics in the short time frame needed—2.5 million years, which is not long in evolutionary terms—to alter something as fundamental as life history characteristics.

A related preadaptation that probably evolved along with prolonged life history characteristics is what appears to be a genetically regulated mother-son incest avoidance programmer (Turner and Maryanski 2005). When chimpanzee females make themselves available to males for sex in what is sometimes termed by field researchers as *the lineup* (where males patiently wait in line for their turn with a receptive female), the already-born sons of these females are conspicuously absent, reducing the chance of inbreeding depression. Just when this incest avoidance pattern evolved cannot be known, but it would give selection something to work on when the ancestors of present-day chimpanzees began to allow sons to remain in the natal community (with only females transferring away from their mothers). Indeed, as we will see, it is this biology-based incest avoidance that would be critical in the transition to the nuclear family among late *Homo erectus* or, perhaps, only early humans.

Mother-Infant Bonds

Virtually all mammals evidence strong social ties between mothers and infants and often strong ties with fathers as well. In the case of great apes, however, paternity could not be known, and presumably male and female transfer patterns away from their mothers at puberty ensured sons would not interbreed with their mothers and that daughters, by chance, would not interbreed with their unknown fathers. The emergence of a biology-based sexual avoidance pattern among chimpanzees appears to be an adaptation of mothers and sons living in the same community. In the mammalian brain, a discrete area—the *anterior cingulate gyrus*—is responsible for many of the unique characteristics of mammals, such as mother–infant/young offspring bonding, separation cries when mother and offspring become separated, and play among young mammals.

Because most mammals organize themselves into some form of kin grouping, infants and mothers enjoy the protection of the group. Such groups did not exist among the last common ancestors to great apes and hominins, and so selection worked on emotions as one strategy to form stronger bonds. Several barriers to forming a kin unit like a nuclear family. First, how were males to be made to commit to females in conjugal pairing up, and further, how were they to be made to bond with the infants born by the female? Second, how were sexual relations between adults in the conjugal pair and their opposite-sexed offspring to be avoided with the arrival of puberty of their offspring? Clearly natural selection was able to install an inbreeding avoidance between mothers and sons, but could the same occur with fathers and daughters? All the data on incest among humans report that father-daughter incest is more common than mother-son and that, when incest occurs, it is always more harmful to the son psychologically than the daughter. I have speculated, along with Alexandra Maryanski (Turner and Maryanski 2005), that crossing both biological avoidance programmers and cultural taboos against incest may be why sexual relations between mothers and sons are less frequent but more harmful when they do occur. It may be, I suspect, that father-daughter incest is only regulated by a cultural taboo and, thus, is not as strong as the combined power of biological programmers *and* cultural taboos against mother-son incest.

Moreover, incest between brothers and sisters is also quite common, compared to mother-son incest, but this is often initiated by males in a more coercive pattern of sexual abuse. Yet, the basic problem in highly promiscuous animals like chimpanzees and their common ancestor with hominins remains: How are individuals sharing so many genes to avoid sexual relations and reproduction with offspring when the transfer patterns evident among apes must be suspended or at least delayed past puberty? With transfer at puberty for males and females among the LCAs of all great apes in the distant past, both offspring move away from their parents—thus ensuring that interbreeding among individuals sharing 50% of their genes is avoided.

Without some means for prohibiting sexual relations among family members, forming the nuclear family of father, mother, and their offspring would potentially reduce fitness among early hominins trying to form more stable kin and group units. If the LCA of contemporary chimpanzees and hominins evidenced the pattern that we see among chimpanzees—mother-son sexual avoidance—then some of the potential problem is resolved by this bioprogrammer. If females transferred at puberty, even with a father in a nuclear family, maintaining this pattern, which is genetically driven, might be sufficient to resolve the problem of incest. Moreover, Edward Westermarck (1891, 1891 [1922], 1926), an early Finnish sociologist, recognized that young humans (and this may extend to other mammals) raised in contact through play activities reveal a pattern of sexual avoidance when sexually mature; and so, what is termed the *Westermarck effect* may also mitigate the problem of potential incest between siblings.

Still, with only the mother–young offspring bonds to build upon, creating a stable nuclear family posed a potentially serious problem of incest, and unless the potential problem could be resolved, hominins could not have survived, and humans would not have come into existence. So, at best, we can speculate that the current sexual avoidance pattern between mother and son chimpanzees was in place among early hominins as well as some form of Westermarck effect. Would these be sufficient to prevent widespread father-daughter and sibling incest and, hence, the genetic deformities of close inbreeding?

A further complication here involves how selection was enhancing emotions, including those surrounding sex, in order to pull males and females into a more permanent bond, as is evidenced by the enlarged *septum* (one of the key centers for sexual pleasure). Selection along these lines could have made males and females sexually aroused for the three dyads of the nuclear family: fathers toward daughters, sons toward mothers, and brothers toward sisters, and vice versa. Emotions can forge strong social bonds, but they were being used to forge strong sexual bonds to create the conjugal pair. Modern-day human families are often rocked by negative emotions revolving around incestual sexual attraction; and such must have been the case before the brains of hominins were sufficiently large to form cultural codes, such as an incest taboo. How, then, did selection get around these problems? An answer to this question is important in understanding the nature of humans. Let me come back to this question a bit later in this chapter, after examining the remaining preadaptations.

Lack of a Harem Pattern in Mating

A harem pattern of mating is quite common. Among monkeys, it is the dominant form. As we saw in the use of monkey societies as a control on the features of great ape societies in Chapter 2, monkeys are group oriented, with females remaining in their natal groups and becoming part of lineal and collateral matrilines of related females. Males in monkey societies are generally driven to migrate away from their natal group to another group and, then, begin competition for dominance in their adopted group, leading to the formation of dominance hierarchies. The dominant males generally seek to maintain an exclusive mating relationship with females, but in fact, because the females are so well organized (and, at times, have their own dominance hierarchies), these efforts at exclusivity often do not work out in actual practice. Yet, it is the effort that counts because, first, it is genetically driven and, second, it precludes other mating patterns, whether the promiscuous pattern of great apes or the more familial pattern (of conjugal parents and off-spring) of many mammals. As noted in the previous chapter, the lack of a harem pattern did not get in the way of selection moving great apes to a more nucleated family composed of a mating pair and their dependent offspring. The great advantage of the monkey harem pattern is that it creates stable structures at the group level, but the existence of a bio-programmer for reckoning and for being oriented to the larger community, with easy movement around the community, was probably incompatible with the high degree of group structure evident in monkey societies.

To add structure to gorilla and great ape societies, for example, natural selection took a different route, strengthening for a time the bond between females with offspring to lead silverbacks until the time female and male offspring transfer away from the mother, thereby often breaking the lead silverback-mother relationship. Yet, this system did provide structure at a critical time during the period of mother-offspring vulnerability. The only other alternative is exhibited with chimpanzees, where sons remained in their mother's community, with all females transferring away from mothers at puberty to new communities, thereby eliminating the likelihood that daughters would have relations with fathers and, at the same time, mixing genes as females moved about communities. Mother-son incest was, as noted earlier, regulated by what appears to be a biology-driven sexual avoidance in chimpanzee communities, facilitated by the visiting patterns in which males visit their mothers but do not form stable groupings with them.

Yet, if more structure was needed among the hominin ancestors of the LCA with the ancestors of extant chimpanzees, then this increased structure had to be built around the moderate-to-strong ties in chimpanzee-like societies: strong bonds between mothers and sons; moderate-to-strong ties among male siblings, and moderate-to-strong ties with non-kin friends. Selection could initially work on these ties to generate more "groupness" that would protect all the members of a community as they moved into more open-country conditions and, in particular, allow mothers to have matings and offspring with other males, except their sons. Then, in order to generate a nuclear family, it was necessary to (1) pull mating partners into conjugal relationships and (2) increase commitments of the mating pair to offspring resulting from their sexual activities, (3) at the same time prevent incest between fathers and daughters (since there were no bioprogrammers to prevent sexual relation, as is the case with mothers and sons), and (4) avoid incest between siblings until the daughter leaves the community at puberty. All this pulling together of family members would be based on charging up positive emotions, including those about sexual relations, the potential for incest and conflict might well increase, thus placing an even greater burden on what was initially a rather fragile nuclear family—just as is the case today among humans in nuclear families. Before speculating further on how this probably occurred and affected human nature, let me briefly outline the last of the preadaptations listed in Box 2.1 on page 45.

Play among Young Mammals

Young mammals play almost universally, with such play involving some rather serious and complex activities, such as to assume a role (e.g., aggressor vs. pursued), to initiate play activities, to coordinate switches in roles (say, becoming an aggressor after being pursued), to be aggressive without hurting play partners, to understand "the rules" of the game, and other interpersonal techniques for coordinating (play) activities. For chimpanzees and other animals that employ rather sophisticated interpersonal practices (see Box 2.2 and the next chapter), it is necessary to learn and practice interpersonal techniques in order to release the bioprogrammers that guide interpersonal behavior; and play when young is critical to the neurological capacity to coordinate interactions without powerful bioprogrammers. In the case of great apes, these play activities allow the young to develop their interpersonal skills for life in a community without permanent grouping and comparatively few strong ties.

For some time, diverse scholars have recognized the importance of play as a necessary stage of development of humans' capacity to engage in more complex interactions that achieve interpersonal attunement (e.g., Mead 1934; Huizinga 1938 [1955]; Beckoff and Pierce 2009; Bellah 2011). Others have stressed that play is necessary to release the neurological potential of the brain and, at the same time, to create and learn cultural codes necessary in interactions with others. As the brain grew among late hominins like *Homo erectus*, play would serve as practice for developing language skills and the ability to assume the perspective of cultural codes.

Play could also work to increase social bonds among siblings, as is evident with male chimpanzee brothers who develop moderate-to-strong ties, and even more so among unrelated males who have played when young and developed stronger friendships with age. Play could have also increased the Westermarck effect between male and female siblings, since this effect appears to be activated by physical contact between opposite-sexed siblings. Thus, a bioprogrammer appears to be present, lessening sexual attraction among opposite-sexed siblings in nuclear families. This bioprogrammer can be activated by physical contact in play before puberty (see Turner and Maryanski 2005: 53–80 for summaries of relevant data). Moreover, play with parents can also work to forge stronger bonds with potentially promiscuous males and with mothers, bringing all members of the emerging nuclear family together through the positive emotions generally aroused by play activities (Bellah 2011: 74–91; Burghardt 2005; Deacon 2009).

The Evolution of the Nuclear Family

We are now ready to put together a scenario about why and, more importantly, how promiscuous descendants of the LCA of great apes and hominins could create the nuclear family, which is the key structure from which hunting and gathering bands were constructed. Thus, much of human nature is connected to how hominins organized reproduction of the species in something not natural to evolving great apes: the nuclear family. Without this critical transformation to hominin societies, humans could not have evolved. How, then, did blind natural selection bring about this most improbable creation—the nuclear family? This basic social structure was not, I believe, part of humans' biological nature. Yet it was nonetheless essential to the survival of all hominins trying to adapt to more open-country terrestrial habitats, at first in Africa and then later in Europe and Asia.

The Primal, Pre-Kinship "Horde"

It may seem a bit odd to evoke an old concept from scholars first speculating in the nineteenth century on how the first groups, and kin groups in particular, evolved (e.g., Bachhofen 1861 [1931], 1967; McLennan 1896; Morgan 1871 [1997], 1877 [1985]). This is the concept of *the horde*, and for many early-twentieth-century scholars, such as Durkheim (1912), there had to be some kind of social formation that preceded highly organized groups like the nuclear family. None of these scholars had the advantage of being able to do cladistic analysis on great apes to make their inferences, but they were, I believe, essentially correct. High levels of group organization were not endemic to the ancestors of humans, and thus, there must have been a "transitional form" in the movement from less to more organized groupings. The term *horde* was often used as a kind of conceptual gloss or filler. It may seem to be an archaic term, but it is rather close to the actual pathway by which hominins carrying the great ape genetic legacy for weak social ties became increasingly organized into group structures.

Today in west Senegal, small populations of chimpanzees spend considerable time on the Africa savanna[1] and retreat to trees scattered across the open-country habitat in order to sleep at night, comparatively safe from most predators. What has emerged among these chimpanzees is somewhat more permanent groupings that perhaps draw upon the genetically controlled propensity of chimpanzee males to engage in patrols of their home range. A result is that mostly males but, - at times, females are involved in-hunting and also scavenging activities. Moreover, these more savanna-dwelling chimpanzees develop new technologies for doing so and even begin to form divisions of labor around male hunting and female gathering.

From Horde to Nuclear Family

Thus, it appears that the initial horde that evolved out of chimpanzee-like community formations of unrelated female immigrants, sons and their mothers, male friends, and brothers could be adapted to quasi hunting and gathering lifeways (with retreat to the forests at night). These tendencies can occur with relatively small-brained chimpanzees who are probably like the first hominins (5.0 to 4.5 mya). If we move forward in evolution to *Homo habilis* (2.2 mya) or early *Homo erectus* (2.0 to 1.8 mya), where selection had already increased the range of emotions for forming social bonds, we can see that the horde that formed likely consisted of stronger ties among (1) brothers born into the community, (2) their male friends, (3) their mothers (protected by the sexual avoidance bioprogrammers from incest), and (4) probably some females who had immigrated into a community that was increasingly trying to

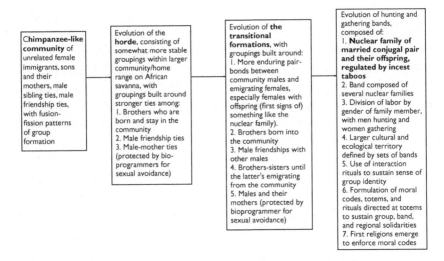

Figure 4.1 The Evolution of the Nuclear Family and Hunting and Gathering Band

organize into more stable groupings. Shortly, or during this initial formation of the horde, sisters of brothers would stay in the group, at least until sisters reached puberty. Figure 4.1 outlines in more graphic form what I see as the sequence of events leading to the emergence of the horde, or precursor to the nuclear family.

As the horde evolved, selection on structures such as the *septum*, the source of pleasure for sex, increased capacities for emotions of love, commitment, loyalty, and so on, that led to increasingly closer relations among sexual partners, mostly incoming females from other communities. There may also have been patterns of relations among males, brothers, and their friends' mothers, but the key was to begin pulling the conjugal couple together. Incest avoidance was probably not complete, but these evolving hominins would soon learn of the consequences of inbreeding (gene-based deformities). At the same time, mother-son incest avoidance and perhaps the Westermarck effect among brothers and sisters who had played in their youth could lower rates of sex between males and their young sisters who had yet to transfer away from their mothers. Moreover, this same Westermarck effect could also have applied to any of a male's brothers and friends if they had engaged in roughhousing play activities with this male's sister as data on humans indicates that play involving physical contact among young males and females who are *not* siblings also activates the Westermarck effect in early adolescence (see Turner and Maryanski 2005 for citations to data;

see also Maryanski, Sanderson, and Russell 2012). As brains grew, perhaps as the very beginnings of speech became evident, culture as a force behind social organization would begin to be used to specify sexual taboos and, moreover, appropriate relations among conjugal partners and their offspring.

We know the end result of this movement through a horde phase of social organization: hunting and gathering bands built from nuclear families (consisting of parents and their offspring), revealing a clear division of labor where men hunt and females gather (often assisted by offspring and, at times, adult males). Once culture begins to evolve, selection can work much more rapidly because intelligent hominins like *Homo erectus* can make rules, the violation of which can lead to activation of powerful emotions like *guilt* and *shame* that can bring direct negative sanctions from others (*anger*, for example), which can then activate *fear* as well. The key was to get sufficiently organized in order to survive and let what natural selection had started—growing subcortical emotion centers to a point where enhanced emotions would make enlarging the neocortex fitness enhancing. Once this process began, language and culture could be used to order social relations, thereby increasing fitness further. And even before *Homo sapiens* evolved, it is clear that this basic societal form was highly adaptive because *Homo erectus*—the last hominin before humans—was able to migrate to Europe, Asia, and down to southeast Asia. Other forms of humans, such as Neanderthals, could also live in diverse and often somewhat harsh environments.

During these monumental transformations from loose-knit communities to more organized hordes and, then, to nuclear families and nomadic bands of hunter-gatherers, selection was working on the preadaptations examined in this and the previous chapter. Rather sophisticated behavioral and interpersonal capacities of hominins carried much of the burden that sustained social relations clearly present in the great apes today (see list in Box 2.2 and the discussion in next chapter). As the range, intensities, and nuance of emotions increased with selection on the subcortex of the hominin brain, the brain began to grow and open the door to language and culture. As a result of interpersonal skills that all great apes and, hence, all members of the horde possessed grew more powerful, allowing late hominins to overcome the barriers to the nuclear family. As a consequence, the great ape and then horde structure could be transformed into the nomadic bands composed of a half dozen or more nuclear families wandering a territory in a nomadic and seasonal route around a home range. Other bands would do the same.

Bands sharing language and culture might also share territories, but it is more likely that they had somewhat different nomadic routes within a larger home range. All this was possible, of course, because humans sustained their orientation to a larger community and to larger horizons beyond the group. The bioprogrammer to reckon community was thus not wiped away but was instead supplemented by a larger palette of affective states that increased the abilities of animals with high levels of interpersonal acuity to form stronger social bonds. Even as descendants of the LCA separating hominins from the ancestors of chimpanzees began to form nuclear families and bands, it was not direct bioprogrammers doing to heavy lifting. Rather it was interpersonal skills charged up by new emotions and growing cognitive capacities that allowed later hominins to form new kinds of flexible social structures that were only *indirectly* related to their biology as natural selection enhanced emotions and, in so doing, opened the door to a larger brain, spoken language, and culture. The result was a basic structure—the hunting and gathering band—that lasted for hundreds of thousands of years, but this basic structure did not lock humans into small-scale, group-level organization as is the case with other primates. About 12,000 to 10,000 years ago, the evolution of mega societies of many millions of persons would begin. Thus, even as humans became more group oriented, they never lost their great ape capacity to visualize a universe beyond simple groupings—for great apes, community—but, for humans, complex layers of ever-larger societies and intersocietal systems. Groups allowed for survival, particularly the nuclear family, but they did not put on blinders to more extended forms of social organization.

Conclusions

For natural selection to drive evolution it must have something to select on. If survival rests on the formation of groups, and especially kin-based groups, it needs variations within the genome, expressed in the distribution of phenotypes, that would move populations of what were, in essence, great apes in brain and body size. As emphasized in the previous chapter, the two most important were (1) a palette of basic primary emotions and (2) a neurological system in the neocortex substantially prewired for language, if sign-based or speech-based languages would facilitate group formation. The preadaptations examined in this chapter do not, in any immediately obvious way, facilitate tighter-knit groups among weak-tie evolving great apes. Community as the basic unit of organization and concern with the larger boundaries and demography of membership are part of great ape nature, especially for the LCA of the ancestors of contemporary chimpanzees and humans' hominin ancestors. Selection on these would not move a population dramatically to a more monkeylike pattern of group formation, although the patrol activities of community males around their community's boundaries was one possible opening for selection to expand this one persistently formed group in chimpanzee communities. Indeed, since it appears to be how present-day chimpanzees in west Senegal create foraging groups in open-country habitats, this kind of grouping was probably an initial point of selection on hominins millions of years ago. But the chimpanzees in west Senegal exist in the most marginal of ways, and we should remember that no other subpopulation of chimpanzees has adopted this pattern of organization in more open-country habitats.

The cognitive mapping of the community, and low levels of inter-personal grooming among great apes, does appear to be a node on which natural selection could go to work. However, as we will see in the next chapter, the tendency of chimpanzees to engage in carnival—emotional ceremonies apparently marking a community—could be a target of selection, and perhaps the cognitive mapping propensities of the LCA to chimps and hominins could be selected on and extended to smaller group formations, giving them a ritual basis for affirmation and solidarity at the group level of organization.

Mother-infant bonding clearly provides a point on which selection could go to work to prolong bonds across generations, perhaps subverting not only male transfer, as is the case for chimpanzees and perhaps hominins' last common ancestor with the ancestors of chimpanzees, but also female transfer, at least for a time, in order to form a more permanent group. The propensity of chimpanzee males to form moderate-to-strong relations with brothers and with non-kin male friends was surely selected on, if this trait was evident in the LCA to chimps and humans. The moderate-to-strong relation of sons to their mothers in chimpanzee communities would thus be the point on which selection could work *if* the LCA of present-day chimpanzees and hominins was programmed with this behavioral trait. Indeed, I have argued that the original horde probably consisted of sons, their male relatives and friends, and at least one or two of their mothers with dependent offspring (Turner et al. 2018: 137–143).

The promiscuity of male-female sex and the lack of strong ties between most adult males and females in the community was the major obstacle to using mother-infant bonds as one-half of a nuclear family, which needed a stable father. The lack of a harem system of mating left open the one and only avenue for blind natural selection to randomly hit upon: *enhancing longer-term bonds between sexual partners to create the nuclear family*. Thus, the lack of a harem pattern of organization among the ancestors of present-day great apes and past hominins can be seen as a preadaptation for the nuclear family. Emotions as they forged stronger bonds, especially by enhancing emotions associated with sex, could start to pull male and female sexual partners into more stable relationships. Then, as emotions allowed for brain growth later in hominin evolution and eventually to spoken language and culture, relations could be enshrined in cultural beliefs and norms and celebrated with ritual activities already in the hominin repertoire of behavioral propensities (see next chapter). Life history characteristics would make bigger brains (and the capacities for speech and the culture that they would bring) possible. A stable family would allow a longer period of infancy for offspring who had a larger but incompletely developed neocortex to move through the female cervix. The longer juvenile period would sustain female care of the still immature offspring up to puberty. As a result, selection only had to work on the neocortex, rather than the complex anatomy surrounding the female cervix (which, in any case, the pelvis and hip joints can accommodate only so much change if they are not to impede bipedal walking that evolved first and was, therefore, critical to early hominin survival). Indeed, without life history characteristics that are simply part of the ape genetic legacy, the brain of hominins probably could not have grown to human proportions and, as a result, neither could language and culture.

Play was essential to hone the needed interpersonal skills of adult hominins, especially skills in managing emotions. Play was also important, I would hypothesize, in activating the Westermarck effect that could supplement the mother-son sexual avoidance pattern evident in chimpanzees and, hence, probably in the LCA of chimpanzees and hominins. Enhancing play by increasing the emotional payoffs of relationships and, also, by generating at least stop-gap measures to reduce incest and its consequences were probably critical in making the horde more fitness enhancing than it would otherwise have been.

It is probably important to emphasize that the emergence of humans depended on a fortunate confluence of events, where selection was able to use preadaptations to begin the process of making hominins more group oriented, not so much by installing strong bioprogrammers for groups (which would be difficult) but by indirect means via the enhancement of emotions and, then, language as it generated cultural norms and other systems of moral codes. During this process, the emotions of *guilt* and *shame* evolved because they put teeth in sanctions and self-appraisals about conformity to expectations inhering in cultural codes.

One consequence of the absence of hardwired traits for group formations was that natural selection had to work around this deficiency and discover a different route (i.e., enhanced emotions to brain growth to language to culture). This route ensured that the structures that did emerge would be *highly flexible*. They could be adapted to diverse habitats, as is evident in the migrations of late *Homo erectus* and early humanlike forms. They would make the construction of mega societies not only possible but, as we will explore, more in tune with humans' evolved nature than were early patterns of horticulture and agrarianism. This flexibility relies on the sophisticated behavioral and interpersonal capacities of humans—capacities were inherited from the LCAs of the ancestors of great apes and humans. Even though these capacities were subject to selection, it may be that by simply energizing these capacities with more intense, diverse, and nuanced emotions, selection may not have had to alter them in significant ways. Indeed, by enhancing emotions, selection set into motion a direction of evolution that led to bigger brains, more cognitive capacities to deal with others, language, and culture—all of which made the interpersonal capacities of hominins more powerful without having to subject them to direct selection. Thus, much of what I can confidently proclaim to be "human nature" is, in reality, great ape nature, *supercharged by emotions, bigger brains, language, and culture.*

Note

1 See, for relevant data: Baldwin (1979); Baldwin et al. (1982); McGrew (1981), (1983), (1992), (2010); McGrew, Baldwin and Tutin (1981); Tutin, McGrew, and Baldwin (1982); Pruetz and Bertolani (2007); Stanford (1990); Mitani and Watts (2004); Mitani and Rodman (1979); Hunt and McGrew (2002); Pruetz (2006); Moore, Bangergraber, and Vigilant (2015); Hernaqndez-Aguilar, Moore, and Pikering (2007); and Langergraber et al. (2011).

5

Interpersonal Skills for Species Survival

Without bioprogrammers directing individuals to specify particular forms of social organization, except for the larger community home range, how were selection pressures for increased organization at *the level of the group* to be realized? As already emphasized, enhancing emotions was *the* breakthrough; and with more emotions, the hominin descendants of the LCAs of great apes developed stronger ties and, eventually, the necessary structures—nuclear families and bands engaged in hunting and gathering economic activity—that would allow hominins on the human clade to survive for hundreds of thousands of years in increasingly diverse habitats.

Emotions alone would not be enough to create more stable group structures because the majority of primary emotions and variants and elaboration of these emotions are negative in the sense that they do not promote social solidarity. Most mammals have bioprogrammers directing organization that can override negative emotions and periodic episodes of conflict among group members, preventing conflict from breaking up the groupings necessary for survival. But for a weak-tie animal like the hominin, descendants of the ancestors of the great apes, increasing the number of positive emotions and charging these up also meant charging up the negative emotions that could disrupt solidarities and thus destroy necessary kin and group structures of evolving hominins.

Even as brains got larger, leading to spoken language and cultural codes, these attributes would have little power to regulate individuals *unless backed up by emotions*. As outlined briefly in Chapter 2 and in Table 3.4 on page 76, one solution was for natural selection to combine (in an unknown way) the three negative primary emotions—*fear, anger,* and *sadness*—into the emotions of social control: *guilt* and *shame* (Scheff 1988; Turner 2000, 2007). While the pain of these two emotions is sufficiently powerful to prompt individuals to engage in behaviors that lead to the avoidance of *guilt* and *shame*, they alone cannot have held groups together. They too are negative and often activate defense mechanisms that create new tensions within an individual and between individuals. If only *guilt* and *shame* operated, then, a world of neurotics and perhaps psychotics would prevail, making stable groups almost impossible to form and sustain over time. To some extent, elaborations of positive emotions outlined in Chapter 3 could compensate for the effects of negative emotions on solidarity, but something more was needed.

This "something" was the *behavioral and interpersonal capacities* that, it seems, were hardwired into great apes and their descendants: *hominins* on the clade to humans. In the end, using expanded capacities to experience and express great varieties of more nuanced emotions (see Tables 3.2–3.3 on pages 62 and 66), coupled with the inherited behavioral and interpersonal capacities listed in Box 2.2 on page 49, was enough to bring the conjugal pair together into more permanent social bonds and to maintain commitments to care for offspring in small bands of nomadic hunter-gatherers (Chapais 2008). If great apes had not been so interpersonally skilled, then humans would probably have never evolved, and instead, their bones would be buried as fossils in open-country habitats in Africa with no one to dig them up. This chapter provides a review of the inherited interpersonal capacities that constitute a most important part of humans' basic nature.

Inherited Capacities for Interaction and Solidarity

Early Imitation of Facial Gestures Revealing Emotions

Sequences of development in the life cycle often reflect evolutionary sequences, and such is the case with emotions. Human infants can imitate within weeks of birth their caretaker's facial expressions of primary emotions (Emde 1962; Ekman 1984; Sherwood 2007; Sherwood et al. 2005, 2008; Tomonaga 1999; Subiaul 2007; Horowitz 2003; Gergely and Csibra 2006). A smile can be returned by an infant, as can a frown and other gestures signaling negative emotions. Thus, human infants are *programmed to read emotions many months before* they can understand and emit vocal gestures, indicating that this developmental sequence of emotional signaling *before* talk probably reflects evolutionary sequences in which a nonverbal language of emotions evolved long before speech. In fact, it is likely that the late capacities of hominins for a more auditory language probably piggybacked onto the neurological wiring for the nonverbal language, or *the language of emotions* mentioned in Chapter 2 (see also Turner 1999, 1997a, 1997b, 2000, 2007, 2013; see also Clark 2012). Moreover, the discovery of mirror neurons (Rizzolatti et al. 2002; Rizzolatti and Sinigalia 2008; Ferrari and Rizzolatti 2015) in primates demonstrates that the imitation of facial gestures is a hardwired propensity, thus giving natural selection a genetically driven behavioral capacity on which to select that which would enhance learning of emotional gestures and the implicit "language" they contain.

It was probably not necessary for natural selection to even enhance this initial orientation of infants to reading and communicating emotions. The propensity to imitate emotions is clearly built into the hominoid genome; and once present, it begins the process of bonding infants to caretakers, intensifying the mother-infant bond and, perhaps later in hominin evolution, a father-infant bond. As more nuanced types of positive emotions were mutually read and used by the young and their parents to form attachments, the emotional basis of the nuclear family began to fall into place. Thus, the nuclear family is not a *direct result* of a genetically driven bioprogrammer that is part of human nature[1] but, rather, a derived consequence of wiring for emotional communications, beginning with the exit from the womb and ending near death. *Being emotionally attuned and responsive* is the critical part of humans' biological nature and, in the end, is the force creating and sustaining solidarities and group bonds.

Reading of Face and Eyes

Great apes are able to read gestures in the face and eyes of conspecifics (Osgood 1966; Menzel 1971; Turner and Maryanski 2008). In fact, they will follow the gaze and eye movements of others, revealing a clear propensity to role-take with others and determine what they are thinking, feeling, and preparing to do (Hare, Call, and Tomassello 2001; Call 2006; Povinelli 2000; Povinelli and Eddy 1997; Itakura 1996; Baizer et al. 2007; Tomasello, Hare, and Fogelman 2001; Okomot et al. 2002).

Visual dominance in mammals, which is comparatively rare compared to the number of species that are olfactory dominant, allows male mammals to visually read gestures of others in face and eyes as the main means for communicating. True, vocal calls can communicate specific states, such as danger, but without a more complex vocal apparatus, great apes cannot communicate more complex meanings in sequences of vocalization, as is the case with a full-blown spoken language. They can, as we now know from research on the language capacities of great apes (Rumbaugh 2013, 2015; Rumbaugh and Savage-Rumbaugh 1990), signal complex meanings via their dominant sense modality—vision. These capacities and propensities for visual communication are hardwired, and because emotions are best read via vision (as opposed to touch, smell, or even hearing), natural selection took this direction: enhance those areas of the body best able to communicate emotions (eyes and face). The result is an animal that can use and understand a large palette of emotions by simply reading eyes, face, and body countenance long before the descendants of this animal—humans—would begin to rely on speech as a form of communication. Yet, speech does not eliminate humans' need to understand emotional undertones revealed through eyes, face, and body gesturing.

Thus, humans still rely on reading the gestures of eyes and face for emotional cues even with a well-developed auditory system of speech. Speech can be emotional, but people almost always cross-check talk with what is being communicated by eyes and face (and bodily countenance). If the two diverge, people will generally trust the eyes and face, if they are trying to interpret emotional meanings. Speech is thus a wonderful means of communicating instrumental intentions and so much else, including emotional dispositions, but it is still the primal language of emotions that humans use when they really want to understand what another is feeling. Of course, this propensity can be used for manipulative purposes, as is the case when someone is conning another for personal gain by seeming to express "authentic emotions" in eyes and face, thus giving "the line" for nefarious purposes credibility (Goffman 1959). It is also, I suspect, why serious poker players hide their faces with baseball caps pulled down, dark glasses, and facial hair. It is not easy to be "poker faced" because controlling facial expressions is difficult when emotions are aroused; it is easier to cover them up, to the degree that would be allowed (poker players wearing masks would probably be prohibited, for example). Expressive control is thus difficult, which it is probably a good thing to keep interactions from turning into "con games."

Capacities for Role-Taking and Empathizing

Great apes can read nonverbal gestures of conspecifics and humans to determine their mental states and intentions and, then, respond by engaging in appropriate behaviors. De Waal (1996, 2009, 2016) has argued persuasively that great apes can empathize with others, read their emotions, experience the emotion in relation to self, and then respond in an appropriate way. A famous recording shows Kanzi, a bonobo chimpanzee, learning English by simple immersion among English-speaking trainers (Rumbaugh 2013, 2015; Savage-Rumbaugh et al. 1978, 1988, 1993, 1994). Kanzi was trying to help his sister, who had not learned English, as the trainer asked her to do something with her arms. Naturally, having not learned English like Kanzi, she could not understand, and her gestures reflected her concern and frustration. Kanzi's gestures did the same, and then he started to manipulate his sister's arms and hands to meet the request of the trainer. This is, in reality, a complex activity: to read the spoken gestures of the trainer and understand what is being requested, to read his sister's nonverbal gestures and emotions in face and eyes to experience empathy with her emotional state, to recognize that he needed to help his sister, and then go to her rescue by grabbing and attempting to move her arms in the correct way so as to meet the trainer's request.

Premack and Woodriff (1978) thus asked whether chimpanzees have theory of mind (TOM), or the ability to know what others are thinking and likely to do and then engage in appropriate behaviors. Many others (Hare, Call, and Tomasello 2001; Povinelli and Vonk 2003; Povinelli, Nelson, and Boysen 1990: Mitchell 2011a, 2011b; Meltzoff 2002, Parr et al. 2005) have likewise argued for this theory of mind—which has become the way that biologists talk about a behavioral capacity initially termed *role-taking* by the pragmatist philosopher George Herbert Mead (1934). Mead used the term in his social psychology class in the first decade of the twentieth century. Because biologists do not read much, if any, sociology and perhaps philosophy as well, they apparently have been unaware of the entire pragmatist tradition that outlies what Mead termed *role-taking*, by which he meant the capacity to read the gestures of others, verbal and nonverbal, in order to "place oneself in the minds" of these others and assess the role or roles that others are trying to enact and then coordinate actions to facilitate cooperation with these others. Whatever the label, chimpanzees and the other two great apes have this capacity, as do a few other highly intelligent animals such as dolphins and elephants as well as some birds. Mead did not believe such is the case, but it is now clear that higher mammals and a few species of birds can role-take, or exercise their theory of mind.

This ability to role-take is the essence of interactions among humans. This capacity is greatly enhanced by the ability to read the gestures of others for emotions because it is the best guide to determining the likely actions of others in group contexts. The ability to read emotions, anticipate how they will affect another's behaviors, and then to make adjustments to these likely behaviors is what allows humans to cooperate. It is the key ability of great apes in communities because encounters are not so much driven by genetically programmed behaviors but, instead, by a much more generalized capacity to read, interpret, and respond to the perceived internal states of others, thereby giving interaction a considerable amount of depth and flexibility. As the variety and nuance of emotions that could be read in the gestures of others expanded during hominin evolution, the subtlety and complexity of interactions could increase considerably. Most importantly, the ability for hominins to "plug into" each other's emotional states not only facilitated the flow of interaction with others, but it also would lead to bonding when positive emotions were mutually communicated. Thus, by simply adding more emotional content to role-taking, the mechanism by which hominins interact also became a bonding mechanism, which would be critical in establishing group solidarities and many elements of what Nicholas Christakis has denoted as the "social suite" of cooperation in societies (see Chapter 1).

It might be tempting to say that forming strong bonds is part of humans' basic nature, and such might be the case, but the low-sociality primates from which we descended did not create such bonds, beyond perhaps those evident among chimpanzees (male friendships with each other, brothers, and mothers). It might be, I think, more accurate to say that humans have in their nature the inherited ability (if activated by play) to role-take and to assess the expanded emotional states of others. It is the ability to (1) role-take and (2) experience more emotions than other animals that is a key element of humans' biological nature. Bonding is an outcome of these behavioral capacities, although many social relationships and groups fall apart because negative emotions can easily and rapidly be aroused in human encounters. Thus, the basic biological traits—role-taking and emotionality—can be, as they often are in complex societies, double-edged swords that promote bonds of solidarity or, alternatively, break them apart. Which side of the sword is operating is largely determined by additional interpersonal propensities to fall into rhythmic synchronization, experience collective effervescence, and engage in ritualized behaviors.

Rhythmic Synchronization of Interaction

The discovery, initially with monkeys, of mirror neurons led to a search for such neurons in great apes and humans. A neurological basis for role-taking, empathy, and mimicry is clearly evident in the type of neuron that mirrors the neurons in others and generates particular patterns of gesturing (Rizzolatti and Sinigalia 2008; Rizzolatti et al. 2002). Great apes do not, however, have the same levels of rhythmic synchronization of talk, body movements, and nonverbal gesturing as humans, most likely because they have fewer bundles of mirror neurons. Yet, mirror neurons and the capacities that they allowed among the LCAs of great apes and hominins meant that natural selection had something to select on—tails of bell curves of those hominins with larger numbers of mirror neurons—and, in so doing, directional selection could enhance role-taking and empathizing in ways that would promote emotional solidarity.

Many sociologists (e.g., Goffman 1967; Collins 2004; Turner 2002, 2007) have documented that human interactions fall into patterns of synchronization of talk and body movements, generating a kind of back-and-forth rhythm. In fact, when interactions lack this synchronization, interactions fall "out of rhythm" and, as a result, are difficult to sustain. Indeed, rhythm is critical to sustaining the flow of positive emotions, and when rhythm is disrupted, negative emotions are aroused in an interaction.

Great apes do not reveal quite this level of, nor even this human dependence on, rhythmic synchronization, although young great apes in their play activities reveal considerable rhythm and synchronization of responses back and forth. Among adults there is not as much synchronization, although arousal of emotions will lead to mimicry of the emotions, which creates a kind of mutual rhythm back and forth. Part of the explanation for this difference among apes compared to humans probably resides in a fewer number of mirror neurons and, perhaps more importantly, in their inability to engage in finely tuned articulated speech, which, among humans, drives much synchronization in what ethnomethodologists in sociology describe as "turn-taking" in conversations (Sachs et al. 1974), as will be explored further in later chapters. Moreover, when turn-taking is converted to chants or to poetic and musical hymns, talk generates even more rhythmic synchronization of bodies and emotional states (Collins 2004). Thus, spoken language may be necessary for the enhanced levels of rhythmic synchronization that humans display over chimpanzees. Still, the basic capacity and propensity were there to select on during hominin evolution, although the long evolution from a language of emotions to one based on speech may have been as important as any direct selection on distributions of mirror neurons.

Collective Emotional Effervescence

Since the nineteenth century, studies of free-ranging chimpanzees have documented that when larger numbers periodically assemble in propinquity, they begin to display behaviors much like those displayed in human "carnivals." In 1844, for example, Thomas Savage (Reynolds 1965: 157) described an assembly of "not less than 50 engaged in hooting, screaming and drumming with sticks on old logs." In 1896, R. L. Garner, a self-trained zoologist, observed chimpanzees in Gabor and described a sequence of events where chimpanzees made a drum-like object out of clay obtained from the bank of a stream and then put it on a bed of peat in order to increase the resonance. When the drum was ready, Garner (1896: 59–60) described a kind of "riotous gambol" or carnival:

The chimpanzees assemble by night in great numbers and the carnival beings. One or two will beat violently on the dry clay (of the drum), while others jump up and down in a wild and grotesque manner. Some of them utter long, rolling sounds, as if trying to sing. When one tires of beating the drum, another relieves him, and the festivities continue in this fashion for hours.

Vernon Reynolds studying wild chimpanzees in Uganda witnessed something rather similar whereby chimpanzees would use tree planks to create a "big drum" sound by beating their hands and feet on specific planks, along with calling, hooting, and screaming. Assemblages from different parts of the forest appeared to move toward each other and, when they met up, went into a "wildly excited state" (Reynolds 1965: 156–159).

These acts of carnival among humans recorded by Spencer and Gillen (1899) in their famous work *Native Tribes of Central Australia* were immortalized in Emile Durkheim's (1912) portrayal as "collective effervescence," the fundamental basis of ritual solidarity among humans (as well as in Freud's analysis in his *Totem and Taboo*). The effervescence, when experienced collectively, was for Durkheim the origin of the sacred and supernatural that typify religions. Thus, humans clearly inherited a behavioral propensity to engage in collective carnivals where emotions are charged up and, as a result, generate an intense sense of solidarity (Maryanski 2019).

Chimpanzees also can display less animated movement of bodies in co-presence. Jane Goodall, who has spent more than five decades studying chimpanzees (Goodall 2005), describes what she terms an emotional *rainfall dance* where chimpanzees stand near a roaring waterfall, seeming transfixed by its power and beauty, swaying in rhythmic movements and displaying a very humanlike wonder. Thus, in this kind of quiet rhythmic synchronization where lower-key emotions are aroused, it is apparent that chimpanzees can engage arousals of emotions that seem almost like "religious worship."

More contemporary theorists have documented that every day, normal interactions among humans can have this sense of effervescence, but at lower levels of intensity, that build social bonds and solidarity. Interactions that fall into a rhythmic flow and synchronization can thus generate less intense emotions; and as these lower-key episodes of emotional animation are repeated over time in subsequent encounters, they cause a growing sense of collective solidarity (Goffman 1959, 1967; Collins 1975, 2004; Turner 1988, 2002, 2007). Thus, the mechanism for charging up emotions, in both high-key and low-key modes, was part of our inheritance from the ancestors of great apes, certainly the LCA of chimpanzees and humans. Once the range of humans' emotions expanded, this mechanism could be used not just in big collective gatherings but *in almost all kinds of interaction where ritualized greetings, synchronization of talk, and emotional arousal at many different levels of intensity occur.* Thus, in expanding the emotional repertoire of hominins, a bioprogrammer for what Durkheim terms *collective effervescence* during episodes of co-presence and interaction could be used to charge up emotions and, later, when culture began to evolve, to charge up the emotions attached to turn-taking, synchronizing of speech, and the cultural symbols and codes generated by the ability to speak.

Seeing Self as an Object

As the pragmatist philosophers of the late-nineteenth and early-twentieth centuries emphasized, humans have what scholars like George Herbert Mead (1934) thought were a unique capacity: to see themselves as an object in their environment. This capacity to see self, as it turns out, is not unique to humans but still rare among animals. The mirror test is one way to assess whether an animal can see itself (Gallup 1970, 1979, 1982; Gallup et al. 2014). This test simply places a mirror in front of an animal, and then observations are made as to whether this animal can recognize its own reflection. A dog, when shown an image of itself, is likely to bark at its own reflection as another dog or go around to the back of the mirror to sniff for olfactory cues to learn more about this reflection. In contrast, a human child will recognize itself, moving and preening to see its movements reflected in the mirror. Similarly, members of the dolphin family will immediately recognize themselves and start to "clown around," apparently having a good time. (Videos are readily available online of dolphins/mirror tests.) Elephants also can recognize themselves; and I suspect if one could find a big enough mirror to lower into the water, so would whales. Moreover, I suspect that some species of birds, such as intelligent parrots and macaws, have a sense of self and thus can see themselves in the mirror.

Charles Horton Cooley (1902) recognized that when individuals interact and role-take with each other, the gestures of others represent a "looking glass" or mirror in which a person sees him- or herself reflected. These gestures often involve judgments and evaluations by others of an individual. If the self-images reflected back from gestures of others are supportive, the individual will experience a quiet *pride,* and when they are not supportive, the person will feel a sense of low-level *shame.* Thus, the capacity of humans to see themselves as objects in their environment and as objects of evaluation by others gives focus to what individuals learn from others in role-taking. Humans seek not just clues about how others are to behave but also information about the self that they are trying to project in the interaction. Thus, interaction is a mutual assessment of dispositions of others and self-conceptions or identities that are driving their respective behaviors. For humans, the processes are vastly more complex than in great apes and, hence, humans' early hominin ancestors, but the basic process of mutual presentations of self and reading these self-presentations is a capacity inherited from hominins and used as humans tried to get better organized.

Part of humans' basic nature, then, is to see and evaluate self vis-à-vis others' evaluations of self and in reference to collective moral codes in the culture of a society. As emotional capacities of hominins expanded, and later as culture began to evolve, self could be seen and evaluated with many diverse emotions, both positive and negative, and from different cultural standards. With the expansion of the scope of emotions and growth in the scale of societies and their cultures, humans could see themselves from many different angles: as a whole person, as a member of salient categories, as a member of particular types of corporate units, or as an incumbent in status positions requiring particular role performances within corporate units. The *fundamental mechanism* remains the same as it probably was with the LCAs, however: individuals see themselves in the "looking glass" of others' responses and, most particularly, the emotional reactions of others. As the brain grew and allowed for more memories to form and to be retrieved, these emotional reactions to the responses of others accumulated over a person's biography and affected the ways that this person acted and responded to others. In this way, the emotions exchanged in interaction become part of individuals and their fundamental psychology. Humans inherited these mechanisms that revolve around emotional imprinting part of great ape and then hominin neurology. But the unique aspects of human neurology (e.g., a large set of nuclei in the subcortex, a very large neocortex, a capacity for speech and culture) dramatically expanded the degree to which emotions could penetrate and regulate each human. Still, even as self—as a mechanism distributing emotional inputs—became more complex and robust, the basic processes remained much the same as they were among the first hominins and a core part of human nature.

Reciprocity and Calculations of Justice

Exchange and Reciprocity

Higher primates, including monkeys, appear to be hardwired for reciprocity—that is, individuals needing to give back to those who have bestowed benefits on them (Cosmides 1989; de Waal 1989, 1991, 1996; de Waal and Brosnan 2006). Indeed, many mammals in general also appear to have a sense of reciprocity, as was the case when my cat would kill a rat (good) or bird (sad), leaving it at my front door in an apparent exchange, I presume, for my feeding him regularly. As emotions were enhanced and as a sense of self became a constant object of evaluation, this deeply embedded bioprogrammer for exchange reciprocity took on even more significance. Exchange, per se, increases the arousal of positive emotions, as documented by a number of experiments using human subjects (e.g., Lawler, Thye, and Yoon 2009 for a review and theory). Once positive emotions are aroused, they become an additional and often more valuable resource on top of whatever else is being exchanged. A present is reciprocated not just because of the rewards inherent in the present but also because reciprocity by others is highly rewarding, in and of itself, as it arouses positive emotions about self and positive emotions about the gift giver. Again, a sense of self focuses and intensifies the emotions involved because a present is given to *a person*, and the emotions thereby aroused signal to a person the gift giver's positive evaluation of sense of self. When gift giving is reciprocated, the positive evaluations become mutual and generate stronger social bonds and, in the end, are one of the core mechanisms by which solidarity among humans is achieved. Of course, the opposite is also possible: failure to reciprocate a gift invites negative emotions and, thereby, reduces social bonds and solidarity.

Exchange and reciprocity thus work to strengthen bonds, even among monkeys who are already hardwired to form strong kinship and group bonds. For a weak-tie, lower-sociality, less-prone-to-group-formation animal like a great ape, the lack of strong bonding and grouping bioprogrammers places a greater burden on reciprocation of exchanges of resources while adding additional levels of positive emotions to reciprocated exchanges. A sense of self increases this sense that reciprocation *must occur* because conceptions of self and identities can be on the line. As culture enshrines in codes the morality of reciprocation, exchanges become a central mechanism for forming and sustaining bonds and solidarities. Thus, there is not, I believe, powerful bioprogrammers for strong social ties, outside of mother-infant bonding among humans but rather neurology-based biological traits such as emotions, self, and reciprocity—all inherited from our hominin and great ape ancestors and all intensified because selection on human neuroanatomy operates almost like direct bioprogrammers on exchange relationships. The only direct programmer is the apparent higher mammalian need to reciprocate, which, among humans, is intensified by emotions, self, and cultural rules that demand reciprocation, thus ratcheting up positive emotions and achieving what was necessary during the evolution of hominins: increasing strength of social ties and group solidarities.

Calculations of Justice

Both monkeys and apes, as well as many other mammals, calculate justice and the perceived fairness of exchange relations. For instance, a new-world capuchin monkey will stop exchanging with a caregiver if another capuchin monkey is receiving more rewards for emitting the same behaviors asked by caregivers (Brosnan et al. 2005; Brosnan and de Waal 2003; Brosnan 2006). Rather complicated calculations are involved in this behavior, including an assessment of the level of rewards to one monkey in relation to other monkeys in the same exchange relationship, perhaps a negative emotional reaction to differences in the rate of exchange from caregivers, and then a refusal to exchange any further until rates of exchange of behaviors for treats are equal. This too seems to be hardwired, and if there was ever a hardwiring for morality in general, this apparent emotional sense of "injustice" is certainly one source.

Chimpanzees do much the same thing. For example, a recent study of chimpanzees documents a rather altruistic sense of morality, above and beyond the capuchin who is concerned only with itself. A chimpanzee would cease exchanging with a caregiver because a relative (e.g., mother, sister, or brother) was not receiving the same level of reward from the caregiver. This behavior involves some extra steps, one of which is the chimpanzee who ceased exchanging probably role-took with his relatives, imagined their emotions, and was upset (experiencing negative emotions), thus leading to a termination of the exchange until equity was restored. Here, there is not just a concern for the deprivations to self but also a sense of fairness for others and perhaps for the larger community in which the chimp has lived (Lents 2016). Again, we are seeing something here like a hardwired sense of morality, which, as the brain grew among hominins, could in fact be enshrined in highly moralized cultural codes. Coupled with the evolution of the capacity of humans to experience *shame* and *guilt*, the *anger* associated when exchanges are unfair provides a hardwired basis for the formation of the capacity to be moral, in the sense that self and others are evaluated by reference to moral codes that specify good and bad, right and proper.

In a more naturalistic setting, male chimpanzees have been seen to hunt for meat (Fahy et al. 2013), with the kill being divided up with a kind of implicit moral formula: the male most responsible for the kill would get the most meat with the rest of the meat being distributed in proportion to the perceived contribution of each individual to the kill. This calculation is thus a more complicated calculation of justice; and it is likely that the first hunter-gatherers used this same neurology-based formula because it was so pervasive among those hunter-gatherers originally studied by anthropologists. Moreover, this calculation of justice involves seeing each individual as a distinctive entity, with evaluations of their perceived contribution to a collective task determining their worth to the collective task and, hence, their deserved share of the kill.

As the brains of early hominins, which were about the same size as contemporary chimpanzees (400 cc), began to grow, these calculations of justice would increasingly pull in more complex and nuanced emotions, more explicit layered senses of self, more active role-taking, and eventually, more specific moral codes sustaining the biology-based impulse for proportional distributions of resources in relation to the respective contributions of members engaged in a collective activity. The underling emotions and the focus that self and identity give to emotions would thus increase the power of the bioprogrammers that pervade higher mammals and particularly primates.

Yet, moral codes could also subvert the bioprogrammers. Studies of exchange where it is not entirely clear who has made the most contribution to a collective task have found that a norm of equity is invoked: distributing the rewards equally among all members. Even at times when there is clarity about respective contributions, these considerations are ignored so that equality in distributions can promote solidarity without differentiating who is more deserving. The apparent reason for this kind of norm is that equity reduces negative emotions, at least among a larger number of members, and thus sustains the solidarity of the group. Haggling over who did what and who deserves what is, among humans, almost always a process laden with negative emotions; and thus, a general cultural norm can be established to distribute equally to avoid disagreement about distributions. For weak-tie animals like humans, who at their genetic core are still a great ape, this norm of equity has probably proven useful in avoiding conflict that would easily destroy solidarity.

A dramatic event, reported some time ago in national newspapers, can perhaps throw some light on this issue. Older chimpanzees are too strong and contentious to remain with their trainers or, if raised in a home, with their caretakers. A couple who had raised a male chimp in their home brought a cake to the compound where he was being cared for and went over to a table and sat down with their "child" to enjoy the cake. One or more of the other males in the area became quite upset over their fellow "inmate" getting such a treat while they got nothing. One male was so enraged that he went over and attacked the "father" to such an extent that his life was threatened, and even if he survived, he would be permanently disabled. This anger was over a real sense of "injustice" when one of his "roommates" received something that, in his mind apparently, should be equally shared but was not. Such is the power of feelings of injustice, not only among older, contentious (but sill incredibly powerful) chimps who could tear a NFL linebacker apart if enraged. With all the enhancement of emotions among humans, the capacity for *rage* can be compounded by additional dangerous emotions like *vengeance* and *hatred*. The arousal of negative emotions over perceived "injustices" is thus the other side of "justice" and reciprocity; yet, for hominins to become more strongly bonded and group oriented, it was probably necessary for justice calculations to be normatively regulated so that ambiguities over respective contributions and distributions of rewards would invoke a default norm: distribute equally when unsure about how to avoid the negative emotions aroused over perceived injustices. Such would be particularly likely in animals with few bioprogrammers for forming social bonds but with supercharged emotional capacities.

The Capacity to Make Attributions

Related to self and calculations of justice is another behavioral capacity evident among chimpanzees and, hence, humans' hominin ancestor: the ability to make causal attributions about the "causes" or sources of experiences. When animals can have a sense of self, they tend to make causal attributions for why events occur and have consequences for self. Attribution theory in psychology (see Weiner 1986) examines these dynamics somewhat differently from my portrayal here. Attributions by humans can be either related to the actions of self (that is, the person caused a particular outcome to self) or externalized to other objects in the environment: others, the local situation, or more remote structures (corporate units, categoric units, or whole society). For example, when a student gets a poor grade in a college exam, the options are to blame self (lack of studying), others (the professor, TA), a corporate unit (school), a categoric unit (ethnicity, religion of protagonists), an institutional system (higher education), or society (and its inequalities and injustices). Defense mechanisms to protect self often make full *self-attributions* ("it was my fault") less likely than *external attributions* ("it was the fault of others, situation, social category, or social structure"). Chimpanzees also possess this human capacity (Kaneko and Tomonaga 2011).

Attributions involve some extra cognitive steps to, first of all, identify classes of objects and others in the environment, which chimpanzees can do. They can make rudimentary assessments of who is responsible for a particular experience. As the brain began to grow among later hominins, this capacity to classify would expand to more others and objects in the environment. Subsequently, the emotional reactions, including defensive reactions, toward self or external others and objects would also increase. In a series of experiments, Lawler and colleagues (2009) developed a theory on the types of exchange that lead to either external attributions to others and the group or to self as responsible for particular outcomes. A self-serving cognitive and emotional bias prompts self-attributions when a person has positive emotional experiences that appear to be hardwired as part of the basic attribution dynamic, so, many attributions among individuals are self-attributions. Because this imperative for greater social solidarity among members of a group or even members of a society would require external attributions to the group, society or some intervening structure had to evolve if hunting and gathering bands and then groups in more complex sociocultural formations were to remain relatively stable and cohesive. Society could not possibly survive if humans were only a bunch of selfish egoists, as some theories (mostly economic) have incorrectly posited.

Thus, Lawler (2001) asked the following: Among the various types of exchanges, which are the most likely to lead to external attributions where the group as a whole would be considered "the cause" of positive experiences? Such external attributions would increase solidarity among group members and commitments to the group—the key imperative on which selection was working to install in hominins. His conclusion was that productive exchanges, where individuals are coordinating their efforts to realize a goal but, at the same time, cannot fully assess their respective contributions, are the most likely to lead to external attributions to the group regardless of whether they are fully successful in realizing their goals. Participants in these kinds of exchanges where they coordinated their labors perceived that "they were in this together" and, moreover, they would experience "a sense of efficacy," both of which are highly rewarding, involving positive emotional valences above and beyond actual success in meeting goals.

The result was that they increased their commitments to the group, even when goals were not realized. These findings suggest what probably occurred in early hunting and gathering societies where coordinated divisions of labor between men and women, and some coordination among men in hunting and women in gathering, was a *productive exchange* in the band. It also suggests a productive exchange in the nuclear family unit as females brought home what they gathered and males brought home their share (as often determined by their contribution) of a collective kill. At both structural levels—nuclear family and band—failure to secure enough food did not lead to blaming others but, instead, to sharing during times of shortages and, moreover, to reaffirming commitments to coordinate activities to secure more favorable outcomes in the future. So the process of beginning to organize into groupings—say the early horde, as portrayed in Chapter 4—did not lead to mutual finger-pointing as others were blamed for failures but, instead, commitments to do better next time. As a consequence, even failure may have strengthened commitments among those engaged in productive exchanges. Thus, under the conditions of organization that were evolving as hominins ventured into open-country habitats, initial groupings increased external attributions to the group rather than to self or specific others and, in so doing, increased commitments to the group. This hardwired propensity to make attributions thus work favorably to push for group formation and productive exchange within family and band in ways that enhanced fitness.

This process of productive exchanges revolving around a division of labor between females and males also has some basis in great ape biology. Chimpanzee males often coordinate hunting for meat in the forests, signaling with subtle nonvocal gestures that are difficult for humans to detect (Menzel 1971). For example, coordinating the hunt of a baby baboon might involve subtly communicated instructions among three chimps about who is to roust the baby baboon to flee, who is to cut of its exits, and who should be in a position to catch it, and where to move off to share the meat among only themselves. Chimpanzees in west Senegal also hunt out on the African savanna and, unlike most other chimpanzee populations, spend considerable time on the savanna in more cohesive groupings than forest-dwelling chimpanzees. Somewhat like hunter-gathering human populations, these males primarily hunt (although at times females are involved as well), while females gather plant life and, as a result, eat much less meat than males (Fahy et al. 2013). Because males hunt so frequently and eat so much more meat than females, evidence from measure of isotopes can distinguish males from females. This division of labor may not be wholly genetically driven but, instead, arises from convenience because males cannot nurse the young and thus are more likely to be free to engage in coordinated open-country hunting, especially since hunting and gathering mothers prolong breastfeeding (as a form of birth control) and infancy of their young. Such was probably also the case for the first hordes as they evolved into the hunting and gathering band, but still, it is likely that there is probably some biological push for this form of division of labor. As such, it can be seen as either a preadaptation or behavioral propensity when collecting food in dangerous habitats. Furthermore, this ability to hunt and thereby gain more protein can also be seen as a preadaptation or behavioral propensity for brain growth. Natural selection would not need to create this division of labor that appears to be a default pattern of organization when chimpanzees hunt. The only change required of human hunters would be for men to increase the share of meat with females, which may have come from expanded positive emotions emanating from their centers for sexual pleasure, like the *septum*. Once the conjugal partners were pair-bonded, more by emotions than genetic bioprogrammers, and once *both* parents were emotionally attached to offspring in a nuclear family (by bioprogrammers for females and emotions for males), sharing of meat would have increased in the emerging nuclear family among late hominins.

Once hominins spent more time in open-country food collection, the coordination of productive activities involved a male-female division of labor and a pooling of gathered plants and shares of hunts (in accordance with justice propensities) in the nuclear family. Emotions were probably the principal force here, but the division of labor that appears to emerge in chimpanzees certainly facilitated coordinated production and distribution of resources, which is the condition that leads to external attributions to the group, whether successful or unsuccessful, that work to increase emotional commitments to the group (Lawler, Thye, and Yoon 2009). Thus, an escalating or self-reinforcing cycle of initial hunting and gathering in the open country leads to some coordination, external attributions to the group, and commitments to the group, all of which lead to more coordination in the division of labor, external attributions, and commitment to more cohesive groups, and so on. There is clearly some biology here, arising from the related bioprogrammers for exchange, calculations of justice in resource distributions, and attributions for group success or failure leading to group commitments. Enhanced emotionality, growth of the brain, emergence of language, use of cultural codes to coordinate productive activity and to distribute resources, and commitments to groups are thus intertwined. Once put into motion, these mingled forces allowed late hominin groups to survive in highly diverse and often difficult habitats.

Fluid and Episodic Hierarchies

Species of monkeys form linear hierarchies of dominance among males and, at times, among females in matrilines of related kin (Cords 2012). Hierarchies and matrilines are thus the building blocks of group structures organizing populations of monkeys (Boehm 1999). In contrast, the LCAs of present-day great apes probably did not reveal hierarchies of power. Yet, contemporary gorillas and, to a lesser extent, chimpanzees reveal some degree of hierarchy. Other apes such as various species of Asian gibbons (who are not great apes) reveal equality between the sexes in what looks like nuclear family units of mother, father, and offspring; it was this ability to create the nuclear family that led to the success of gibbons and other non-great apes like siamangs. Again, *organization* was the key to survival, but the ancestors of great apes did not possess bioprogrammers for much of this needed property.

So, why did gorillas and chimpanzees develop elements of hierarchy? Both species of great apes are semiterrestrial, with the gorilla being mostly terrestrial because of their large size and the chimpanzee being somewhat more arboreal because they always sleep in the forest canopy at night. Being exposed on the ground in forests, however, is still dangerous, given the potential for predation among animals with weakened olfactory capabilities(after the rewiring of the primate brain for visual dominance) for detecting danger. Among gorillas, the lead silverback is the center of rather fluid groupings composed of males and females typically with offspring. Females stay if the lead silverback is providing services, such as protection and babysitting, but will leave to another group if she is, apparently, dissatisfied with the services provided by the lead silverback. The lead silverback has power in the group, but it is a power that cannot be pushed too far because any member of the group can leave at any time, and indeed, males and females often wander in and then leave the group. So, a hierarchy provides a center around which fluid comings and goings occur, and thus some degree of group organization continuity. It is not a hierarchy, however, that involves *high degrees of control* beyond the requirement to follow the lead silverback around, if they decide to remain in the lead silverback's group.

Among chimpanzees, males sometimes compete for dominance. For example, brothers may support each other for some degree of dominance, but again, the problem is, Whom do they dominate? Highly mobile chimpanzees can leave situations where there is too much dominance. Therefore, dominance does not lead to stable groups, although the patrols of community boundaries may be an occasion when dominance has some utility in defending the boundaries of the community from outside males. So, again, some degree of structure is created by dominance, but it is a weak and often only episodic dominance. Most importantly, it does not ensure continuity of the groups that might temporarily form in some chimpanzee societies.

Since hunting and gathering populations typically revealed a powerful norm against anyone claiming power or prestige, it is likely that this norm was created because bands where individuals tried to appear "above others" dissolved due to conflict that would break simple hunting and gathering societies apart. So, this weak propensity to hierarchy probably evolved from the LCAs of the ancestors of gorillas (8–9 mya) and chimpanzees (5–6 mya), but the stability of the band among hominin hunter-gatherers apparently required that it be repressed. One consequence of these tendencies, rarely expressed, was to increase hominin sensitivity to *status differences*—whether honor and prestige given to accomplished individuals or power given to individuals to control the actions of others for short periods of time. Just as all great apes can role-take (theory of mind) and read each other's minds and dispositions to act, they can also *status-take* and determine status differences. This ability would be important as human societies became more complex and when, as a result, power and prestige were bestowed on some individuals as part of a system of social control, beginning with Big Men among settled hunter-gatherers (often of fishers near oceans, lakes, and waterways), where populations grew significantly and required some form of authority to coordinate the larger population. In these early manifestations of consolidated power, the Big Man often "owned" everything but, in fact, was required to redistribute it to members of the settled community—thereby gaining prestige to go along with his power (in some ways, the Big Man had to redistribute because many of the resources involved were food resources that would spoil if hoarded). Much like the lead silverback among gorillas, power incurred obligations to others and the group that would later in societal evolution be transmuted into a means to exploit members of a society and increase inequalities and stratification. Still, human societies would eventually create complex status systems within and between corporate units in larger and more complex societies, and this ability to reckon status and live with status differences indicates a preadaptation and a behavioral capacity for developing diverse types and fluid hierarchies in societies. When the hierarchies became too rigid, they would, over the long run, generate internal conflict—as can be seen just about everywhere in the world today.

Finally, some people assert that hierarchy and dominance are programmed into humans, whether the argument is extreme like Ardry's (1966) and Wilson's (2019) view of warfare among hunter-gatherers as the driving force of societal evolution or more constrained like Christopher Boehm's arguments about hierarchy in the forests. In

Boehm's (1999) argument about the chimpanzee of Gombe, and also in early work by de Waal (1982) in the colony of chimpanzees in the Arnhem zoo, more hierarchy is "observed" than actually occurs in chimpanzee communities in their natural habitat. In Boehm's case, competition for dominance ensued when the caretakers at Gombe (Jane Goodall's famous site) began feeding members of the community, with the consequence that they began to concentrate at feeding locations and, then, to fight over the distribution of food offered by staff. In de Waal's work, he studied chimpanzees in zoos, which would be much like studying the nature of humans in prisons, where hierarchy and conflict rule to a greater degree than outside prison walls. One response to scarce resources among humans is to fight over them and become dominant so as to control access to resources, but this does not necessarily mean that humans are driven by some diffuse and overbearing need to dominate. It may be the case, but my view is that humans did not survive by dominance; just the opposite, they survived by imposing equality among individuals and by norms against dominance. It is only later when power was needed to control larger populations that we see hierarchies form in human societies, which are more the result of sociocultural selection pressures for consolidating power for control and for coordinating actions among members of much larger and settled populations than of some primal drive for individual power. The case for or against a biology-based "need for power or dominance" is still ambiguous, but the drive for domination appears to be a weak one.

Friendship and Fellowship Behaviors

When the social ties of great apes are examined by gender, one pattern is clear: except for mother-offspring bonds, which are broken for both male and female offspring in orangutans and gorillas and for only females among chimpanzees, *strong social ties of any sort are rare* (Maryanski and Turner 1992; Turner and Maryanski 2005, 2008). The ones that do appear stronger are mostly males' social ties with their brothers or specially chosen male friends (Mitani et al. 2000; Lukas et al. 2005). Occasionally, there are apparent friendships between adult males and females, as is the case when a male orangutan stays for a time with the female that has been impregnated, when lead sliverbacks and females with children form a bond of convenience, or when male and female chimpanzees "hang out" together more than is typically the case. Among females, *all* social ties are weak (except for mother-daughter ties that are terminated at puberty), primarily because the females in a community have all transferred in from other communities to replace those adolescent females who have trans-ferred out (thereby providing a fitness-enhancing "mixing" of genes). The result is that fully adult females are likely to be relative strangers to each other. So when the primal "horde" began to form in the evolution of the nuclear family, selection worked on the male bonds of fellowship with each other and their mothers whose ties may have been hardwired before the split of the LCAs of present-day chimpanzees and those hominins on the human line (Turner et al. 2018; Maryanski 2019).

It is difficult to know if there are differences in human males' and fe-males' propensities for affiliation and fellowship with like-sexed in-dividuals, but such could be the case—although saying so in any kind of definitive way is fraught with conflation with current ideological move-ments and cultural traditions. Among hunter-gatherers, which are our best look at human societies without the conflation of complex modern cul-tural beliefs and social structures, males do form more fellowships, but it is also evident that females do as well, especially with kindred. Thus, again is not easy to isolate a clear biology-based bioprogrammer for friends and fellowships. It may be that, like conjugal couples, these fellowships are built more indirectly from the arousal of positive emotions emerging from interactions. Yet, we should not assume that that humans are "naturally social" because, at our great ape core, we are still an evolved great ape. Old bioprogrammers are not necessarily selected out of the genome but rather are supplemented by new ones. In the case of big-brained humans, cultural norms and beliefs were rapidly supplementing and perhaps even overwhelming bioprogrammers.

Conclusion

This chapter completes the review of what can be derived from a cladistic analysis of humans' closest relatives: the great apes. With the assembled data from cladistics, we have now reviewed (1) preadaptations and (2) behavioral capacities and propensities that were available for natural selection to work on. We know the direction natural selection took: to increase the strength of social ties and group solidarities by, eventually, increasing the strength of emotional social ties between mating adult males and females and between this conjugal pair and their offspring to form the nuclear family. Then, with some stability in the nuclear family, the hunting and gathering band composed of a number of nuclear families could evolve.

The hunting and gathering band composed of several nuclear families proved to be a viable adaptation for hundreds of thousands of years. It weathered the periodic near extinction of humans in the distant past in various locations. As a result, the nearly 10 billion people on earth are all descendants of *tiny* breeding populations that make human primates the least genetically diverse of all primates today.

The fact that humans, at their genetic core, are less diverse than other primates would suggest that the biological traits that I am seeking to isolate *are the same,* or closely so, for all humans on earth. Humans today are descendants of a relatively small breeding population, although superficial differences such as skin pigmentation, eye folds, and shapes of skulls have evolved because humans have adapted to so many diverse habitats. Even noticeable phenotypical differences like skull and facial features or size and height are not controlled by a large number of genes and alleles. In contrast, the genetic basis for behavioral capacities and propensities or brain functions are controlled by many more genes and involve a much larger proportion of the human genome. They are what drive human nature, as I have conceptualized it, and do not vary so easily by point mutations such as skin pigmentation because they regulate complex organic systems where mutations are almost universally harmful. Most of human nature is controlled by complex neurological systems that have not dramatically changed since the emergence of humans. And even variants on *Homo sapiens,* like Neanderthals, were not a separate species because they interbred with *Homo sapiens.* Where differences existed, they were dramatically reduced by inbreeding in places *Homo sapiens* were more numerous, thus leaving only a small proportion of genes—less than 4%—evident in the most Neanderthal-like human today.

This lower level of genetic diversity, then, is explained by bottlenecks in the evolution of *Homo sapiens* and interbreeding among subspecies of early humans. The most dramatic bottleneck appears to have occurred in central/ southern Africa where, by mitochondrial DNA, the genetic diversity in the human population may have dropped dramatically as the number of *Homo sapiens* declined to small numbers, although these data only allow inferences about females. Other studies have found a significant decrease in Africa at one point in humans' evolutionary history. Loss of diversity in DNA in later humans increases with distance from Africa, but there does not appear to have been a large species-wide decrease in genetic diversity like the one that occurred in Africa before humans migrated to other parts of the world. By reasonable methodological procedures, then, it is clear that human genetic diversity did decline at a key point in Africa and then through perhaps smaller bottlenecks outside of Africa (see Amos and Hoffman 2010 and Manica et al. 2007 for summaries of data and references cited). The key point to emphasize here in closing this chapter is that all contemporary humans are descendants from a comparatively small breeding population in

Africa—perhaps only a few thousand individuals and, as some argue, maybe even fewer. For our purposes, this "origin story" suggests that at the genetic level, particularly in our neurology, humans are highly similar, despite surface differences of skin color, body size, skull configurations, eye folds, and other minor traits regulated by relatively few genes and alleles. Thus, *we should all have much the same nature as biological beings*—a nature inherited from the LCAs of hominins and the ancestors of great apes who themselves, I suspect, were small breeding populations, as they are today, based on the lack of fossils on their ancestors.

In the next chapter, I will argue that human emotions and cognition as generated by neurological and body systems are much the same, and they distribute themselves in similar bell curves across all populations in the world. As a consequence, it is reasonable to assume that *elaborations* on the inherited legacy from the LCAs of hominins and ancestors of great apes would have similar effects on all humans as the emotion centers of the brain grew, then as the neocortex grew, and as spoken language and culture became central features of all humans on earth. For what we seek in a book on human nature is to get behind the potentially distorting effects of local cultures and varieties in social structures of diverse populations in order to discover *what all humans have in common.*

It is still difficult to establish what is biological from what is cultural, but we are at least starting with a clear picture of humans' biological heritage. Inevitably though, the fact that natural selection pushed for enlargement of the hominin brain, first subcortically and later neocortically, which in turn led to spoken language and the capacity to form symbolic culture, still makes it difficult to disentangle biological from sociocultural. Moreover, enlarging the brain also changed *the biology* of those preadaptations and behavioral capacities/propensities that we now know to have a biology basis. The next chapter begins the protracted process of examining what, for a better name, I will call *elaborated properties of human nature,* generated by the particular path that natural selection took in enlarging the human brain. Selection, in doing so, generated some new traits that might be considered additions to human nature.

Note

1 There is, of course, some biological basis for half of the nuclear family because female humans like virtually all mammals have a bioprogrammer, probably lodged in the anterior cingulate gyrus (see Figure 3.1 on page 60), for attachment to offspring. The rest of the nuclear family (fathers) is, I believe, generated by emotions surrounding sex, which evolved as a way to create an indirect biological basis for the nuclear family.

6

The Elaboration of Humans' Inherited Nature

The Vulnerability of Hominins and Early Humans

One of the most fascinating mysteries of great apes today is the lack of fossils on their ancestors. The genetic data can be used to establish rough dates for when each of the great apes shared an ancestor with the ancestors of humans (see Figure 2.1 on page 31), but actual fossils on these ancestors of present-day apes are scarce. At 24 million years ago (mya), apes dominated the arboreal habitats of Africa. Later, in various periods from 18 to 10 mya, the forests receded in Africa with climate change, causing some species of monkeys and apes to migrate to Eurasia, where it was warmer. As Eurasia cooled and Africa grew warmer, some of these species migrated back to Africa, but by 10 mya, apes were clearly in decline in both Africa and Eurasia, with species of monkeys proliferating.

Thus, from 24 to 10 mya, monkeys increasingly became dominant, and apes were pushed to the more marginal habitats of the forest or onto more terrestrial habitats such as woodlands, secondary forests, bushlands, and savanna. The fossils, dating back 8–9 mya, may be early hominins (because they were somewhat bipedal) and preceded *Australopithecines* at 5.0 mya but, again, the mystery—where are the ancestors of present-day great apes to be found?—remains. This lack of fossils perhaps indicate there were *never many species* of apes at 10 mya, especially compared to monkeys, which means there simply are not many fossils to find. Indeed, the long-term trend has been such that, today, only 13 species of apes, including humans, have been identified. The majority of these are species and sub-species of gibbons and siamangs, who are smaller Asian apes that are well adapted to the forest (although their numbers are declining because of the human invasion of their habitats). In contrast, we know of some 131 species of old-world and new-world monkeys and even 62 species of Prosimians (or pre-monkeys). Thus, over the past few millions of years, apes in general have been in decline; and it appears that the only successful one has been humans and their late hominin ancestors. Figure 6.1 reports on the relative numbers of species of prosimians (pre-monkeys), monkeys, and great apes (including humans) now in existence.

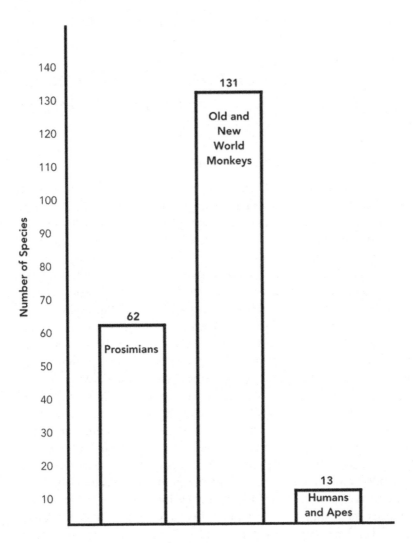

Figure 6.1 Relative Numbers of Extant Primates

The episodes of climate change and, then, the ascendancy of monkeys caused a decline in the number of species of apes, and it is likely that great apes in Asia and Africa will continue to decline because of human incursions. Thus, the suite of adaptive traits were initially successful in the arboreal habitats of Africa, but later these traits became less adaptive because of competition with species of monkeys. These same traits, however, were what was available for natural selection to "select on" as the forests in Africa began to shrink, thereby forcing some species of apes to adapt to more terrestrial habitats. Perhaps surprisingly, many of these traits constituting "ape nature" would become highly adaptive for late hominins and humans and, hence, are still part of human nature.

As mentioned in the conclusion to the previous chapter, humans have come up against near extinction in the relatively short history (in evolutionary time) of *Homo sapiens* (Amos and Hoffman 2010; Manica et al. 2007). The genetic evidence indicates that humans went almost extinct in Africa and only avoided extinction by finding routes out of Africa to Eurasia; now the descendants of this small seed population, which may have numbered only a few thousand, is currently overpopulating the earth. Thus, even as humans became able to form nuclear families and bands in simple hunting and gathering, they were always vulnerable—despite their large brains, language, and culture. Indeed, using those larger brains to create cultures and social systems that are disruptive to the global ecosystem has, once again, made humans and just about all animals vulnerable to extinction. Thus, large brains, language, culture, and even large societies *are no guarantee of fitness* in the long run.

In the past, bigger brains, language, and culture did not insulate human populations against potential extinction because the store of cultural knowledge and the levels of technology were low and because a nomadic hunting and gathering form of social organization left humans exposed to the vagaries of ecological change. Even though the nature of humans as inherited from the last common ancestors (LCAs) of great apes and hominins enabled *Homo sapiens* to survive by getting sufficiently organized, the human population did not begin to grow dramatically until 15,000 years ago. At present, patterns of human organization threaten virtually all life on earth.

The issue to be addressed in this chapter is, What were the features of human nature that allowed early humans to get organized and, later, to expand so dramatically the scale of social organization to the point where humans are once again vulnerable to a degraded environment, this time of their own making? The answer to this question, I think, resides in what I will term *elaborations* on the traits inherited from the LCAs of contemporary great apes and hominins. These elaborations came with evolution of the biology of hominins and humans for enhanced emotions, expanded cognitive capacities, articulated speech, and symbol-based culture. Once these aspects evolved, they would intensify and expand the features of human nature inherited from the LCAs of humans' great ape and hominin ancestors.

Human nature can thus be viewed as a series of preadaptations and behavioral capacities/propensities inherited by the LCAs of great apes and hominins (see Boxes 2.1 and 2.2 on pages 45 and 49) that have a clear biological basis and, as a consequence, were available targets for natural selection to enhance. Natural selection is a blind process that works mostly on existing distributions of traits of phenotypes and the underlying genotypes generating these phenotypes and secondarily on mutations, gene flow, and genetic drift that affect the distributions of traits on which natural selection works. When animals' environments change, selection is often set into motion if existing phenotypes and biologically driven behaviors are no longer adaptive to the new ecology in which a species seeks to survive and reproduce. The emotional capacities of hominins and humans initially evolved in response to selection pressures for stronger social ties and group organization, but once these emotional enhancements allowed the neocortical portions of the hominin brain to grow, preadaptations in hominin neurology for language could be subject to selection for articulated speech, *if* speech would increase fitness. Once an animal can engage in speech, it can begin to accumulate and pass down cultural knowledge, ideas, moral codes, technologies, and other symbolic systems of information that then begin to supplement genetically driven tendencies to behave and organize in particular ways. Human emotion, thought, action, and organization thus began to be driven as much by cultural as genetic codes. And, once articulated speech, expanded subcortical and neocortical portions of the brain, and culture existed among late hominins and early humans, they would operate to *intensify all other preadaptations and behavioral propensities* discussed in the preceding chapters.

Separating the Biological from Cultural

Thus far, through the magic of cladistic analysis, I have outlined the organizational patterns as well as the preadaptations and behavioral capacities of the LCAs for the evolutionary lines leading to present-day great apes and humans. Because of their genetic closeness to humans, great apes provide a good proxy for what the hominins leading to *Homo sapiens* were like. By comparing great ape and human brains, we can get even more data on what natural selection was doing to the hominin brain over the last 2 million years, and hence, a better sense for how the traits inherited from the LCAs of humans and great apes were altered by human neuroanatomy. Moreover, because great-ape habitats under the forest canopy have remained more stable than those of the hominins that created the human line (which would begin to move out from the forest), we can be confident that most of the differences between great apes and humans today are the result of selection pressures on hominins being forced to live in more open-country habitats. True, gorillas and chimpanzees are perhaps more terrestrial than their common ancestors, but they have remained in the forests, whereas eventually *Homo habilis* and, certainly, *Homo erectus* spent increasing amounts of time in open-country habitats where increased organization at the group level would be fitness enhancing. Indeed, the major selection pressures driving hominin evolution were, first, to become bipedal and, then, to become increasingly organized at *the level of the group.* Bipedalism was relatively easy for natural selection to effect because all great apes can stand and walk on their hind legs (as, no doubt, could their LCAs). The real challenge for natural selection was to convert weak-tie, non-group-organizing primates into stronger-tie animals capable of forming more permanent groups with higher levels of solidarity than is evident among great apes today (and yesterday in the distant evolutionary past).

What we have accomplished thus far, then, is some degree of precision in determining the biological traits of the LCAs of great apes and humans at a biological level. As Boxes 2.1 and 2.2 on pages 45 and 49 delineate, the LCAs of humans and great apes had a series of preadaptations waiting to be selected on and an inventory of behavioral capacities and propensities that also could be selected on to make hominins more social and, even more importantly, group oriented. These preadaptations and behavioral propensities represent, I believe, our best look at hominin nature. By isolating capacities and propensities that are clearly based on the biology inherited from LCAs of hominins and extant great apes, we have a picture of human nature *before* the compounding effects of the larger brains, speech production and use, and symbolic culture that would evolve over the past 1.5 million years of hominin evolution. To understand humans' biological nature, it is necessary to understand the nature of the human line *before* subcortical emotion centers, neocortical cognitive centers, speech, and culture evolved to their present-day proportions. Then, by comparing humans today with hominins, we begin to gain an understanding of how larger brains, speech, and culture altered the biological nature of humans. Human nature, then, is the outcome of what was inherited by hominins from the LCAs of *Hominids* (i.e., apes and humans) when this inheritance was modified by natural selection as it generated more emotions (subcortically), more intelligence (neocortically and subcortically by the increased emotionality of hominins), more communicative capacities through speech and, with articulated speech, more powerful modes of cultural production. The alterations in the brains of hominins over the past 5–6 million years of their evolution that led to the emergence of *Homo sapiens* and related subspecies (e.g., *Neanderthals, Denisovans*) some 450,000 years ago are also part of human nature, even though isolating the biology behind speech and cultural production from the cultural and organization products of speech and culture remains difficult. Because we have a picture of what hominin nature was like *before* speech and culture, and indeed, *before* subcortical and neocortical areas of the brain began to grow and allow for speech and culture, we are in a better position to make reasonable inferences about the ultimate biology of human nature, separated from the cultural and organizational assemblages produced by this biology. There will, of course, always be some uncertainty and ambiguity in fully separating what is cultural/organizational from what is in humans' biological nature, but the following analysis provides less uncertainty and ambiguity than is normally the case—as is evident in the illustrative examples of past efforts to understand human nature outlined in Chapter 1.

Brain Growth, Speech, and Culture as an "Elaboration Machine"

Brain growth in subcortical areas of the hominin brain began early in hominin evolution under increasing selection pressures for hominins to become better organized *at the group level* as they were forced to adapt to terrestrial and open-country habitats—first the ground under the forest canopy, then secondary wooded forests, bushlands, more open-country habitats in Africa (e.g., the savanna), and then migrations by *Homo erectus* or *Homo ergaster* to Eurasia. The brain initially grew at the subcortical level (Turner 2000) because it provided the easiest route to creating stronger social ties and group solidarities among low-sociality and non-permanent-group-forming hominins. Community was the "natural" (hardwired) organizational unit of hominins, but they were increasingly thrust into an environment that required more permanent groups that even contemporary chimpanzees rarely evidence. Initial movement to a more terrestrial habitat led selection to install more groupness among gorillas and chimpanzees than orangutans, who are most like the LCA of all great apes because they still live most of the time high up in the forest canopy like all great apes once did. Moreover, in the only case where contemporary chimpanzees have ventured out onto the African savanna (in West Senegal), they reveal even more stable groups than chimpanzees residing on the forest floor, although even these venturesome chimpanzees move to the sparse trees in this open-country habitat at night.

We know that neocortical growth in the brain did not exceed beyond 425–500 cc for several million years of hominin evolution. Indeed, it is unlikely that until about 1.2 to 1.0 mya, the neocortex went beyond this level, which is not all that much bigger than present-day great ape brains. Subcortical growth must have been well under way, increasing the range, nuance, varieties, and other properties of emotions. We can make this inference because it is now clear that intelligence and emotions are related (Damasio 1994). Intelligence is manifest in enhanced capacities for memory storage, thinking, and decision making, and these capacities depend on emotional tags to cognitions, thoughts, and decisions. Such is the case for *all* intelligent life forms on earth (and I would suspect for all organic life in the universe). Thus, the neocortex could not grow dramatically *until* the emotional palette of hominins had expanded, and dramatically so, perhaps close to the level of present-day humans.

Thus, selection pressures for more permanent and more cohesive groups started the evolution leading to humans. Once the emotional palette increased to a sufficient threshold, selection on growth of the neocortex would be fitness enhancing because a more intelligent animal in predator-riddled open-country habitats is more likely to survive, especially one that could organize for food collection and defense in more tight-knit and stable groups. Human nature, then, is related to *selection as it worked to make weak-tie and non-group-forming hominins more emotional so that they could use emotions to increase the strength of social ties and group solidarities.* The sociological imperative to get organized or die led selection to blindly discover this route to increased strength of social ties and the stability and cohesiveness of groups among rather individualistic apelike animals oriented more to larger communities than to local groupings.

As natural selection grew key nuclei in subcortical areas of the brain, it also set into motion additional selection for growth in the neocortex. Eventually this growth made the rather complex conversion to speech production fitness enhancing. Selection to create the neurology for language, was unnecessary because this basic capacity *already existed* among the LCAs, as is evident by the ability of all great apes to understand spoken language from humans and to communicate back to humans via American Sign Language or via typing on computer keyboards "sentences" linking pictographs carrying common meanings. Even though this complicated process may have taken as much as a million years to install, selection reworked Broca's "hump" or "cap" for downloading speech into Broca's area, along with altering the genes affecting the larynx, tongue, lips, and related muscles necessary for fine-tuned speech production.

Speech changes the world for an animal because it allows for all dimensions of self, thinking, emoting, behaving, and organizing in diverse environments *to be represented symbolically*. It thus becomes possible to create common culture among members of a species, revolving around both idiosyncratic and shared memories, private and public beliefs, shared normative expectations, shared worldviews, codified knowledge that can be passed down across generations, and other cultural products that shape forms of social organization. With the ability to experience emotions such as shame and guilt, cultural prohibitions could evolve, making social control by negative sanctions from others into *self*-control by each individual (Boehm 2013; Turner 2000).

Figure 6.2a diagrams the key relations among growth in subcortical and neocortical areas of the brain as they make speech and cultural production possible. The sequence of this evolution begins, first, with the expansion of emotional capacities within the subcortex that, secondly, leads to not only the growth of the neocortex and but also denser bundles of neurons connecting nuclei and structures within and between subcortex and neocortex. In turn, these alterations in the brain make speech production fitness enhancing, and as speech emerges, so does culture. The more complex speech becomes, the more complex culture can become, especially among animals with brains already loaded with emotional inputs from larger nuclei of the subcortex that increase capacities to remember, to think and reflect, and to make decisions. As the reverse causal arrows in Figure 6.2a, flowing from right to left in the figure illustrate, feedback effects occur in these evolution processes. As neocortical growth increases, its reverse causal effect expands emotional

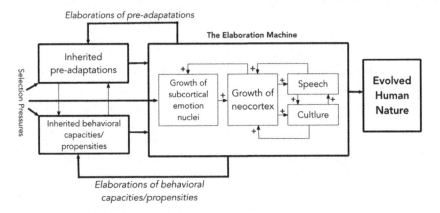

Figure 6.2a Elaboration of Inherited Traits from LCAs of Great Apes and Hominins

capacities. Neocortical growth spawns capacities for speech and then culture, creating reverse effects on neocortical growth and organization, which, in turn, affects subcortical areas of the brain producing emotions. Thus, once selection has started down this path outlined in the direct causal arrows in Figure 6.2a, selection accelerates since each of these new human capacities for emotions, thought, speech, and culture feedback and affect each other. The result is that the biological structures generating emotions, thought, speech, and culture can continue to evolve to the point where the brain cannot be any larger than its ability to pass through the female cervix at birth.

The bold reverse causal lines and arrows to the preadaptations and behavioral capacities and propensities inherited from the LCAs emphasize the powerful effects of enhanced emotions, cognitions, speech, and culture (*the elaboration machine*) on the evolution of hominin nature. The elaboration machine did not, however, strip away the inherited great ape traits of the LCAs. In general, natural selection is a conservative process and only modifies inherited traits or supplements them with new traits in order to increase fitness or the ability to survive and reproduce. It does not try to "maximize" fitness as some models borrowed from neoeconomics by biologists assert; it simply tries to get just above the threshold allowing for fitness. In the case of human nature, the inherited traits from LCAs, as outlined in Boxes 2.1 and 2.2 on pages 45 and 49, remain in the human genome but in *elaborated* form. To make for somewhat overdramatic imagery, these traits of the LCAs are "fed" into an *elaboration machine* with its outputs, as is emphasized on the right side of the Figure 6.2a, being a now *further evolved* human nature.

Figure 6.2b highlights the elements of the elaboration machine from the middle portions of Figure 6.2a. Growth in the subcortex arising from selection to increase social ties and group solidarities allows for the growth of the neocortex because more emotions mean that more complex cognitions can be tagged by more diverse emotions. There may also be some direct selection pressure here when enhancement of emotions puts pressure on the neocortex to grow in order to utilize the power of emotions to expand cognitive capacities. As the neocortex grows, a positive reverse causal effect, and perhaps even a selection effect, is caused by cognitive development to push for new variants and combinations of emotions to tag the increasing development and storage of cognitions. Similarly, an expanding neocortex not only allows for speech but also pushes selection toward speech because it is only through speech that cognitions, particularly cognitions revolving around collectively held

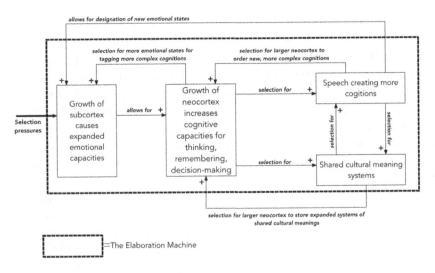

Figure 6.2b Selection Effects among Subcortical and Neocortical Growth, Speech, and Culture

meanings among members of a population, can grow in ways that can increase fitness. Once this cycle of selection is initiated, it continues to ratchet up neocortical growth, up to the point where the brain of infants can still pass through the female cervix at birth.

Spoken language, coupled with a larger neocortex, allows for cultural meanings to be developed and stored as memories that can enhance fitness. Such an elaboration of culture allows for increased memories, eventually written down, among individuals and collectivities *without* having to increase the size of the human brain. Speech probably pushes for culture because, by its very nature, speech links symbols denoting facets of the external environment, interpersonal relations, and intrapersonal cognitions and emotions. Speech also allows for the designation of new emotional states by individuals, both separately and collectively. The development of cultural meanings puts selection pressure on the neocortex (and prefrontal cortex) to expand to store and retrieve shared cultural meanings that can be passed down to subsequent generations. This expansion of the neocortex under pressure to store ever more culture also intensifies the reverse causal effect of a growing neocortex on the growth of the subcortex to produce a greater variety of emotions to tag the increasing volume and diversity of cultural meaning that needs to be stored. Speech, as it makes the formation of symbolic culture possible, increases fitness without enlarging the brain any further at around an average range of

1,250 to 1,350 cc because, eventually, humans acquired the capacity to store culture as myths, poems, and speech enhancers for expanding memories that can be passed down across generations. The invention of writing and eventually printing dramatically expands, outside of the neurology of the human brain, large amounts of information, history, and knowledge in texts and, eventually, computer-assisted systems of memory and memory retrieval. In this sense, a textbook, novel, cloud storage facility, and computer algorithm are all extensions of the human brain.

There is, then, a limit to how far these self-escalating causal and selection effects can continue because there is an upper limit of how big the brain can become, given the structure of female anatomy. Great functional intelligence can be created without growing the actual brain but, instead, giving it the key—speech—to allow for the accumulation of symbolic culture that, in turn, would lead to external extensions of the brain by technologies for storing, retrieving, disseminating, and using information to make decisions or confront adaptive problems. What has always intrigued researchers is why the brains of subspecies of early humans, such as Neanderthals and Denisovans, were so much larger than the human brain, on average, at 1,500–1,600 cc, compared to a modal size of about 1,350 cc for adult *Homo sapiens*. Is this the maximum that will pass through the female cervix? Or was something about the Neanderthal and Denisovan brain different or perhaps less efficient? Answers to questions like this are unclear, but the key point of Figure 6.2b is to emphasize that once selection moves in a particular direction, positive reverse causal effects push selection *to continue movement in this direction* as long as such selection enhances fitness. Once natural selection began to build the elaboration machine as the solution to the organizational problems of weak-tie, non-group-forming hominins, it moved toward growing the brain to the point where speech and culture would increase fitness *to the point* where a bigger-brained newborn could still pass through the female cervix. Once this barrier was reached, humans found new ways to extend the capacities of the human brain with new external information infra-structures. Other restricting dynamics, such as the dangers of making an animal with a larger neocortex too emotional and too smart for its own emotional "peace of mind," may have been in play. Human emotional stability is always problematic, especially as defense me-chanisms and powerful emotions like shame and guilt distort and amplify cognitions and emotions. A majority of human emotions reveal a negative bias, and a population of only neurotics and a sprinkling of psychotics is not likely to be fit. Rather than increase the size of a neurological system already overloaded with emotions, then, natural selection took the route of installing speech production that would allow for cultural production that could eventually be given a boost from writing, text production, computers, and new systems of external memory storage (e.g., the "cloud"), all of which could increase fitness but do not require further growth of subcortical and neocortical areas of the brains (although some might question the fitness-enhancing effects of game technologies, social media, and other such extensions of cognitive and emotional dynamics).

Once selection began the routes outlined in Figures 6.2a and 6.2b, it also elaborated the traits inherited from LCAs of present-day great apes and hominins. When growth in the emotional capacities of hominins reached a threshold level after perhaps as much as 4 million years of evolution, emotional capacities allowed for—indeed, probably pushed natural selection for—a larger neocortex. Then, this change began to accelerate during the last million years of hominin evolution and set into motion the escalating cycles outlined in Figure 6.2b, up to the point where further growth would not be fitness enhancing. Evidence in the fossil record for *Homo erectus* and *Homo naledi* indicate that larger-brained hominin subspecies of *Homo* had brains of widely varying sizes in different habitats and niches in Africa and Eurasia. Some stalled at about 500 cc, whereas others were close to the human measure. Some subspecies of early humans had even larger brains, but the smaller brains of *Homo sapiens* prevailed perhaps by acts of conquest but more likely by simple interbreeding of a larger population (*Homo sapiens*) with smaller populations of Neanderthals (a form of demographic conquest), and so there must be something to the modal size of the human brain, when supercharged by speech and culture, that was fitness enhancing over all the rest.

Human nature today is, then, an outcome of the process whereby the elaboration machine outlined in Figure 6.2b converted the inherited traits of early hominins and the LCAs of extant great apes into more robust complexes of biology-based proclivities to think, act, and interact in particular ways. In trying to summarize this "evolved human nature," I have reverted to portraying this biological nature of humans as a set of *complexes*: (1) the *emotions complex*, (2) the *cognitive complex*, (3) the *psychological complex*, (4) the *interaction complex*, and (5) the *community complex*. Natural selection created these complexes by first selecting on the subcortical portions of the brain to begin the construction of the elaboration machine in order to allow hominins to form strong ties and more stable, cohesive group structures; and once this elaboration machine was fully constructed, human nature at the biological level had significantly evolved beyond the inherited legacy of hominins.

Conclusion: Visualizing the Evolved Nature of Humans

Dramatic presentations of human nature have been offered; for example, many have posited that humans are basically power-seeking and warlike, striving to control others and their territories (e.g., Ardrey 1961, 1966; Morris 1967; Tiger and Fox 1971; van den Berghe 1973, 1975). Even though conflict does occur among great apes, especially among chimpanzees, who can be quite violent under certain circumstances, violence is not a persistent behavioral pattern. Violence does not appear to be persistent among hominins and human hunter-gatherer populations. Humans, under certain conditions, can be violent, as is obvious, but most of the conditions fostering this violence are *sociocultural* in nature. Negative emotions of humans are always present for activation under extreme sociocultural conditions, yet it is doubtful that this violence, even high degrees of aggressiveness, is a central part of humans' biologically driven nature. True, tendencies for some degree of hierarchy among gorillas and chimpanzees may occasion episodes of aggression and even violence, but they do not lead to social control and domination that some posit as the essence of humans. Although hunter-gatherers had powerful norms against inequality—probably because of the dangers of social hierarchies in generating conflict—this normative constraint does not mean that hunter-gatherers were trying to "tame their violent nature." The fact that human societies became more hierarchical once leaving hunting and gathering as a societal form does not mean an aggressive trait in human nature was waiting to suddenly "break through." Rather, consolidation of power and authority simply became necessary for organizing larger, settled populations, with selection for hierarchy being more sociocultural than biological. As human societies evolved rather dramatically over the last 10,000 years, following several hundred thousand years of remaining relatively static, the structure and culture of these more recent sociocultural formations do not mean that they are manifestations of humans' violent nature that can finally express itself. Indeed, I would argue somewhat the opposite—societies reveal how flexible humans are in their ability to adapt to larger, more complex social structures that are *not compatible with their nature*.

One particularly important point in the understanding of human societies is that "kin selection" as conceptualized by biologists does not represent a powerful argument about human nature. Humans' ancestors were *not* kin oriented beyond females' bonding with their young off- spring (a virtual universal among all mammals). *Kinship is a cultural invention* for humans and, as I think the cladistic analysis underscores, the LCAs to humans and great apes were not kin oriented. Indeed, if hominins had not been able to invent kinship by building the first nu- clear family units, humans would never have come into existence. *Overcoming the lack of a drive to form kinship groups* was the big obstacle that natural selection had to overcome, which it did indirectly by en- hancing emotions that strengthened social bonds and group solidarities and allowed for the invention of the nuclear family. Once humans brains could generate a larger variety of emotions, the neocortex also began to grow, eventually allowing for speech and cultural production that, over the last thousand years, led to the mega societies of the present. Without the invention of the nuclear family, however, humans would not exist, and the only mega societies on earth would have been those of insects—which, in the long run, might still come to pass.

The five *complexes* to be examined in the next set of chapters—cognitive, psychological, emotion, interaction, and community—are, as this label *complexes* suggests, more robust and nuanced than most portrayals of human nature, where one or only a few driving forces are highlighted. Moreover, human nature is much more than "driving forces"; it is com- prised of capacities that push and constrain humans to act in certain ways, to be sure, but do not always work in concert with each other. Human nature is often rather contradictory, despite the fact that persons actively seek to sustain a certain congruence among their thoughts, actions, and emotions. Being human with (1) a palette of emotions that can be charged up to destructive extremes, (2) a large neocortex capable of high cognitive powers, (3) a mix of subtle and dramatic psychological need-states, (4) an extensive set of interpersonal capacities, and (5) a capacity to orient to multiple types and layers of social structure and their cultures can all lead humans to experience contradictions, incongruences, imbalances, and other conflicting cognitive and emotional states. Most of the time, however, these intersecting components of our nature work in sufficient harmony to allow humans to adjust to the complex social systems of the modern world. Adapting to simple hunting and gathering societies was relatively easy for humans with this complex nature, but even as societies themselves, driven by their own sociocultural dynamics, became more complex, the five complexes of human nature can be seen as preadaptation to such mega societies. Humans are remarkably adaptable to rather grim social con- ditions—even poverty, crowding, exploitation, inequalities, discrimination, and many other maladies of societies—and yet we continue to reproduce, which, in a biological sense, means that we are "fit"—indeed, the fittest of all the great apes. Being so fit may be humans' undoing as the sociocultural and ecological environment of our own making changes and brings yet more difficult circumstances to which humans must adapt. We need to see, however, just what these complexes involve and, then, to assess whether they can continue to carry the species forward or trigger a decline in the human population.

7

The Evolved Cognitive Complex
and Human Nature

Human nature is greatly affected by the mechanism of memory formation and ordering of cognitions. The processes of how memories are laid down in neuro-nets are fairly well understood, but just how bundles of related memories are ordered for retrieval is not as clear. Thus, a considerable amount of speculation is needed to assess those aspects of cognitive functioning that are a part of human nature. Examination of the *cognitive complex* of human nature comes first because it exerts large effects on all other complexes, even though the emotion and even interaction complexes were well in place before the neocortex began to significantly grow 1.0 million years ago (mya). Of course, the cognitive complex is affected by the other complexes; indeed, the neocortex and prefrontal cortex of hominins would not have grown very much, if at all, without the original expansion of the nuclei in the subcortex generating emotions. Because the emotion complex, as well as the other complexes, was further enhanced after the growth of the neocortex, it is best to start here with the cognitive complex, even though its full dimensions lagged somewhat behind the emotions and interaction complexes that allowed hominins to become better organized in order to survive.

The Nature of Brain Functioning during Action and Interaction

The neocortex of humans is quite large, relative to body size; moreover, the folding of the cortex allows for the packing of additional neuron bundles into the neocortex. We must also remember that the subcortex where emotions are generated is also quite large for humans compared to other primates, controlling for body size that is roughly correlated with brain size. Because the human brain is so large and able to use a wide variety of emotions with which to tag cognitions for memory formation, selection on the brain clearly led to increased density of neuro-nets and integration within and between subcortical and neocortical areas of the brain. Selection initially worked to enhance the brain as speech began to evolve but was constrained by the limitations of speech to organize and remember large quantities of emotionally tagged cognitions. Not until the invention of writing, long after the brain had evolved to its current proportions, was it possible to record information in cultural texts and, thereby, reduce the memory burden on the brain. In fact, as noted in the previous chapter, writing dramatically increased the storage capacity of a population's accumulated information and knowledge beyond that of the average human brain. Still, the brain evolved in ways to increase the capacity of hominins and then humans to remember and to reason in ways that increased fitness.

To do so, natural selection must have created neurological mechanisms that would maximize the storage of information in a world where little could be written down as cultural texts. And because humans today all evolved from a relatively small deme of early *Homo sapiens* from the middle to upper southern portions of Africa, all human brains are probably "programmed" in some way to store, manage, and retrieve information in a symbolic world of speech and emotions without the ability to create elaborated written cultural texts. These "programs" or architectural features of how the brain is constructed or reconstructed, using the template inherited from last common ancestors (LCAs) of hominins and ancestors of current great apes, represent an important part of human nature.

This chapter is thus devoted to the elaborations of the inherited nature of hominins as the brain grew and was rewired to increase capacities for memory storage and retrieval as well as for decision making among animals relying on speech and its capacity to create culture in a universe without writing and elaborate texts. These processes represented an elaboration of the basic properties of the great ape brain and, in so doing, created one of several avenues for further elaboration of the biology inherited from LCAs of hominins and ancestors of extant great apes.

Still Foundational Insights into Cognitive Functioning from Early Theorists

Mind and Thinking

Alfred Schutz (1932) altered the phenomenology project of philosophy to address sociological problems, and the key to this effort was the concept of *stocks of knowledge at hand* that could be rapidly retrieved (by the prefrontal cortex) for use in interpersonal situations. George Herbert Mead (1934) borrowed John Dewey's (1922) conception of thinking in his conception of "mind," which was the behavioral capacity "to imaginatively rehearse alternative" lines of conduct, imagine the outcomes of each alternative, and then select that alternative that would facilitate cooperation with others, thereby increasing the adaptation of a person to a situation. This conceptualization of minded behavior is another way to address the question of rationality, whereby individuals pick alternative lines of conduct that facilitate cooperation in order to bring the most "utility" or "reward" to a person. What Schutz, Mead, and Dewey did not know is that this capacity to weigh covertly in "mind" depends on the emotions aroused (positive or negative) when each alternative is considered in minded deliberations. It would not have even been possible for humans to remember alternatives without the cognitions for each of these alternatives being tagged with emotional valences. Such is the way that organic brains on planet Earth operate, thus making old distinctions between *rationality* versus *emotionality* neurologically incorrect (Damasio 1994). As emphasized earlier, remembering, thinking, and making decisions are only possible with the ability to tag cognitions with emotions; which may, in fact, be the most fundamental dimension of human nature, making humans probably the most cognitive animal on earth and, most assuredly, the most emotional animal.

Significant Symbols, Mind, and Role-Taking

What Mead added was the notion of *role-taking*, which is the basic mechanism that allows individuals to read each other's gestures and, thereby, to "take" the standpoint of others into their minds and then adjust alternative behaviors to facilitate cooperation. Great apes can rather easily role-take, which means natural selection had this basic capacity to work on. As selection first began to expand the subcortex, the new sets of emotions generated by this subcortical growth allowed for the size of the neocortex to increase and be fitness enhancing, in the sense of increased capacities for minded behaviors in Mead's and Dewey's sense of the term or in the sense of economic theories on decision making. As the capacities for mind increased, the evolution of speech created a new means for communicating with conspecifics, although speech was always supplemented by the language of emotions. The emergence of speech valanced with emotions allowed for the articulation and storage of cultural codes in the brains of late hominins or, perhaps, only early humans. The result of these processes creating the "elaboration machine" was to dramatically expand the ability of late hominins to role-take in Mead's sense of the term: to read the verbal and nonverbal gestures of others to determine the dispositions of others to act in certain ways and then, on the basis of this reading of others' minds, to adjust lines of conduct to cooperate with these others in organized, concerted activities. Cooperation in emerging and more stable groups was greatly enhanced by these expanded cognitive capacities for role-taking, which in a more general sense is what primatologists and biologists often term *theory of mind*. When individuals mutually role-take, they "get inside each other's minds" and can thereby more effectively coordinate their activities. This process is influenced by the cognitive capacity to layer self into a series of identities, as will be examined later, where the person is able to understand the identities of others being presented while at the same time presenting his or her own identity or identities salient in the situation to others. This process is part of what Mead meant by role-taking. If we seek to unpack role-taking, it is useful to conceive of individuals as mutually *identity-taking* and *identity-making*, as we will examine in Chapter 10. Therefore, the better an animal becomes at role-taking, the more cooperative actions can be flexibly adjusted, which in turn would increase fitness. Chapter 10 extends this notion of role-taking to not only include identity-taking (of others' identities being presented) but also additional facets of interaction such as the ability to understand others' emotional states (*emotion-taking*), recognize their status (*status-taking*), understand the requirements of situations (*situation-taking*), recognize the social structure to which individuals should be oriented (*structure-taking*), and to be aware of the relevant culture codes (*culture-taking*) for understanding *all* aspects of this expanded view of these additional dimensions that can be added to Mead's conception of role-taking. The notion of "taking on" the role (appropriate emotions, status considerations, reference structures, situational requirements) that others are perceiving and responding to can be examined next to the notion of "making" in the sense that individuals also "role-make" (and by extension, *identity-make, emotion-make, status-make, situation-make, structure-make*, and *culture-make*) by communicating in their speech acts and body language the roles, emotions, status, structures, situations, and culture codes they think or want to guide the interaction. As individuals role-take, they also implicitly or deliberately assert for the duration of an interaction the *role* that they will or want to play (that is, they role-make as they role-take). They do the same for the *status* they seek to occupy or claim, the *emotions* they seek to communicate, the aspects of the *situation* they see as important and the relevant *structures*, as well as the *cultural codes* that should be invoked. Thus, humans have the cognitive capacities to role-take and role-make, to status-take and status-make, situation-take and situation-make, structure-take and structure-make, and culture-take and culture-make. All these are part of the interaction complex, but they are all simultaneously invoked in interactions by virtue of humans' expanded emotional and cognitive capacities. Again, these aspects of humans' fundamental nature will be examined in Chapter 10.

Emotions, Cognitions, and Self

Early theorists, mostly in the philosophical tradition in America known as *pragmatism*, emphasized that the capacities for mind and role-taking (as I have extended them) are the necessary base for the expansion of a sense of self. Great apes have a sense of self, as do a few other animals on earth, but the human self is much more complex. This complexity arises out of the expanded emotional and cognitive capacities as well as the increased interactions among animals that can speak and use culture during interactions. As part of the process of adapting to the social universe, as emphasized by pragmatists, only human interactions involve mutual assessments of others' actions with respect to many dimensions of the sociocultural universe—that is, roles, status, emotions, situations, structures, and cultural codes. Words, voice inflections, and expressive gestures in eyes, face, and body countenance (e.g., the language of emotions) are all in play each and every time humans interact. This process may have been relatively easy to execute in the simple societies of hunter-gatherers, but a considerable amount of fine-tuning was required because humans' emotional, cognitive, psychological, and interpersonal capacities were just as great as they are today. Thus, humans' emotional and cognitive capacities were used, even in relatively easily scripted interactions within just a few basic social structures—nuclear family, hunting groups, gathering groups, and band as a whole. The emotions, psychological needs, preferences, and other dispositions of persons at the moment of interaction still had to be read or taken on, while those of each individual had to be communicated to others. Thus, the interaction was still rather complex, just as all interactions are today in face-to-face contexts. In virtually all situations, individuals are driven by their nature to work through mutual "taking" and "making" of others' dispositions, thoughts, feelings, requests, and other dimensions along which all human interactions flow.

Just as interactions can become more complex, so do the ways that individuals perceive of themselves, compared to the interpersonal activities of great apes (or humans' hominin ancestors). In many ways, the elaboration of self into a series of identities, to be examined in the next chapter on the psychology complex, provides a way to *focus* interpersonal behaviors. Humans all implicitly recognize that individuals are presenting themselves to others by their orchestration of gestures, both verbal and nonverbal. Humans recognize that efforts to understand how another is likely to act depend on understanding the nature of the other's self that is in play. Thus, interaction is influenced by the *cognitive capacity* of humans to layer self into a series of identities, which will be outlined in the next chapter. These identities drive humans to present identities to others for verification. Indeed, humans' cognitive capacity to hold and layer multiple selves during interaction and to orchestrate both subconsciously and consciously gestures to communicate *which* selves, or what are often termed *identities*, are on the line both simplifies and, at the same time, makes for a more complex interaction. Because multiple levels of identity may be on the line, individuals in their identity-taking must work to figure out *which* identities of others are most salient. At the same time, knowing the selves or identities on the line greatly facilitates discovering the roles, status, emotions, situational cues, structural references, and culture that need to be taken into account. Conversely, when a person implicitly knows what identity or identities are important to another or others in a situation, the person's orchestration of gestures to another is simplified, increasing the likelihood that all other "taking" and "making" activities with respect to other dimensions of the interaction will proceed smoothly. A powerful, if not the most powerful, need-state of humans (in the *psychology complex*) is to have their identities verified. Attention to this need generally allows for all other dimensions guiding the interaction (emotions, status, roles, structure, situational cues, and cultural codes) to fall into place because the primary direction of all interactants is to assess the identities and how they can be verified. Thus, the dynamic of identity formation focuses efforts on role-taking and -making and, as a result, makes it more likely that individuals will have a successful interaction

This process of focusing happens swiftly because of the large neo-cortex and rapid-fire prefrontal cortex of humans, which means a great deal of information is being assembled and processed, sometimes without complete cognitive awareness and reflection but simply a "gut feeling" usually based on cues of each other's identity or identities being projected. Much of this argument from early pragmatist philosophy is captured in the label *theory of mind*, developed and expanded by Mead and others in explaining the dynamics of human interaction.

Cognitive Capacities, Self, Emotions, and Defense Mechanisms

For Mead (1938), individuals constantly engage in "acts" revolving around the processes outlined in Figure 7.1. Individuals are driven to achieve a state of equilibrium with their environment, and the processes by which this occurs are, for Mead, "the act." Moreover, humans are typically engaged with different "acts" simultaneously that can be at different stages of unfolding. As a result, humans experience many impulses or states of disequilibrium with the environment simultaneously, and different acts can be at different phases, as outlined in Figure 7.1a

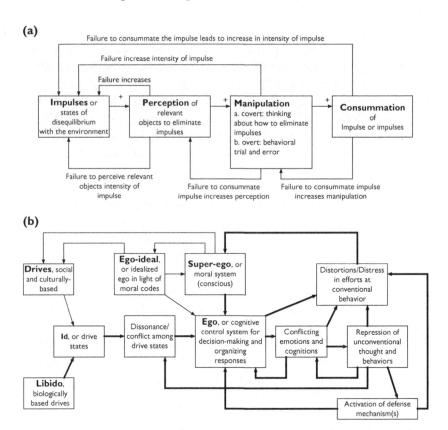

Figure 7.1 (a) George Herbert Mead's Model of the "Act"
(b) Sigmund Freud's Alternative Model to Mead's Overly Cognitive
Model

(i.e., impulse, perception, manipulation, and consummation). An *impulse* is a sense of disequilibrium with some aspect of the environment (hunger, sleep deprivation, tension with others, failure to live up to cultural codes, inability to adequately *role-take*, or failure to verify self). Every person always has multiple impulses in play, with some moving to consummation and others remaining unconsummated. Impulses lead to a *perception* phase, whereby objects in the environment that might potentially be relevant to eliminating or consummating the impulse are perceived. As relevant objects are perceived, individuals move into the *manipulation* phase that can be (a) *covert*, involving minded behaviors where alternative lines of action are assessed for their capacity to consummate an impulse, and/or (b) *overt*, revolving around observable trial-and-error behaviors to consummate an impulse. If these efforts at manipulation fail to consummate the impulse, Mead argued that the impulse will increase in intensity, and especially so if they are self-related impulses, and as a result, the impulse will occupy more and more of a person's perceptual field as they seek ways to eliminate impulses.

In this argument, Mead comes close to Sigmund Freud (1899) because the impulses that are often the most difficult to consummate revolve around the inability to verify self/identities and be at equilibrium between how a person thinks they should be seen and how they are perceived to be seen by others. For example, when a young person seeking parental *approval* or *love* does not receive these sentiments from a parent or parents, and if this failure to receive positive sentiments is chronic, the failure to consummate the impulses generated can lead to a lifetime of failed perceptual and manipulative effort to achieve approval of, or love toward, self.

Mead did not address *defense mechanisms*, and this key idea was, of course, one of Freud's (1899) great contributions to understanding the brain and human interaction. Part of the reason that Mead did not emphasize defense mechanisms is that he did not outline in any detail his ideas on emotions. In contrast, Mead's fellow pragmatist, Charles Horton Cooley (1902), began to incorporate emotions into Mead's basic model. As noted in earlier chapters, Cooley emphasized that individuals experience *shame* or *guilt*, depending on their ability to have their conceptions of self (or identities) verified and accepted by others. Yet, he probably did not go far enough because negative emotions like *shame* are painful and are often repressed, to varying degrees; and once repressed, other *defense mechanisms* delineated in Table 3.5 on page 79 are brought into play. Defense mechanisms may allow a person to limit the emotional pain somewhat, but repression almost guarantees that the underlying impulse will force a person to remain at the *perception* or *manipulation* phases of a particular act. In this state, a person must exert a considerable amount of emotional energy even as conscious awareness of the impulses driving these often compulsive behaviors is long lost to repression. Indeed, if a defense mechanism like *reaction formation* takes hold, the young child not receiving approval from his parents will still repress the *shame, guilt*, and perhaps *anger* and sadness while the same time, expressing great *love* for his "wonderful parents." Thus, the human brain is wired to convolute human cognitions when intense negative emotions are aroused, thus making thoughts and actions often difficult to understand.

Thus, it is part of human nature to (a) seek to cooperate with others because almost all human activity occurs in socially organized that require the deployment of interpersonal skills needed for role-taking, (b) perceive objects in the environment that can consummate impulses, (c) rehearse covertly alternative lines of potential conduct and/or engage in impulsive "trial-and-error" behaviors to consummate an impulse, (d) experience positive and negative emotions with each line of conduct, and (e) select the line of conduct that is most likely to lead to verification of self and meeting others' need-states, thereby gaining the reward of positive emotions about self, others, and situation. Yet, these patterns of action, or "the act" in Mead's terms, are complicated by the following:

- Many such acts are generally in play at any given time.
- Perception and manipulation phases of the act can be unsuccessful and, thereby, increase the intensity of the impulse and activate a new round of perception and manipulation phases of the act.
- Most important acts involve putting self and identities on the line.
- Failure to verify self becomes yet another unconsummated impulse that is emotionally charged.
- Unconsummated acts revolving around self will often activate defense mechanisms that ensure that the perception and manipulation phases of the act will not be successful.
- Acts devoted to verifying self will continue to generate dysfunctional behaviors, arousing more negative emotions that are then repressed in a potential cycle leading to severe behavioral pathologies.

In the larger brain, emotions are *the* essential mechanism whereby cognitions are remembered and decisions are made about how to act and interact. Self and identities become the dominant fulcrum around which thinking and action pivot, and the emotions revolving around self, whether conscious or repressed, will distort thinking, perception, and actions in ways that can be dysfunctional for the individual as well as for cooperation with others. While severe pathologies are often seen as an aberration, they are, in fact, *part of human nature.*

Humans are always on the edge of emotional overload because the central dynamic of brain functioning is emotional arousal for forming memories, retrieving memories, decision making, assessing the reward value of alternative lines of conduct, verifying self and realizing other powerful need-states, dealing with situations where self and responses of others are discordant, and for many other central dynamics to be outlined in this chapter. Animals like humans who, as evolved great apes, do not have powerful bioprogrammers for group-oriented behavior must instead *actively construct* social solidarities. The emotional stakes are *almost always high* and, when things go wrong, difficult to control.

Most of the time when negative emotions arise in interaction, the shame and perhaps guilt aroused by not meeting the expectations of culture or of others in a situation will lead to a redoubling of efforts to receive positive reactions from others. At times, especially when shame and guilt are generated, defense mechanisms will be activated, beginning with repression and then one of the defense mechanisms outlined in Table 3.5. An interesting cognitive capacity, made possible by a big brain trying to sort through cognitions and the emotions attached to them, is to partition these negatively charged emotion-laden cognitions in an effort to maintain a viable self. Of course, repression and defense mechanisms typically create new levels of problems in interactions, but it is a testament to the powers of the prefrontal cortex as it interacts with memory centers (e.g., hippocampus and frontal lobe) to hide from conscious negative experiences and feelings about self. An even more extreme kind of partitioning occurs when individuals reveal "multiple personalities" in which a person brings entirely different selves to an interaction, often not known by one of the other selves, indicating the human cognitive capacity to store and isolate entirely different, often contradictory selves. Typically, one of the selves (or it can even be several selves) are deviant when compared to the more conventional self that the individual also possesses. By separating personalities, cognitively and emotionally, behind partitions in the brain (just how this works in the brain is not known), conscious congruence of conventional identities might be sustained but obviously at the high cost of severe behavioral and emotional pathology. This strange phenomenon of split personalities—an outcome of universal operation of cognitive mechanism and defense mechanisms—is also part of human nature, taken to extremes.

Thus, the theory of mind, as often posited by biologists, evolutionary psychologists, and primatologists is perhaps a useful gloss, but it glosses over too much. Instead of capturing the complexity of cognitive functioning in humans' biological nature, it is only a simple view of the cognitive capacity of primates and humans to read each other's dispositions to act. For humans, however, this process is dramatically extended by the elements of the elaboration machine—enhanced emotionality, enhanced cognitive capacities, speech and true language, and symbolic culture.

Ordering Stocks of Knowledge Used in Interaction

The Emotional Basis of Memory and Experience in Ordering Cognitions

As already emphasized, memory formation revolves around a process of tagging cognitions with emotions generated in subcortical areas of the brain. Cognitions are held in what are termed the *transition cortices* during any interaction (see Figure 3.1 on page 60), thereby allowing individuals to remember what they just said and what others said for the duration of a conversation; the same would be true of any experience as it takes place. The more an interaction or an experience arouses emotions, and the stronger are the emotions aroused, the more likely the emotions will be sustained for the duration of an interaction or for the length of the engagement with the environment generating an experience. If the threshold of emotions is sufficient, the cognition will be tagged with emotions and stored in the hippocampus for a time, and if the cognitions marking an experience are called up as a memory, and the emotions tagging the cognitions will be experienced once again (typically in somewhat diluted fashion compare to the original experience), the memory will continue to be held[1] (Damasio 1994, 2000). The more the emotionally tagged cognition is recalled and the emotions attached to this cognition are re-experienced during a two-year period, the more likely the memory will be shipped from the hippocampus to the frontal lobe for longer-term memory. Emotions, then, are the key to memories in both the shorter and longer term.

Emotions thus determine whether memories will form. If an event, interaction, or experience is not exceptional and, hence, does not arouse emotions, it is unlikely to be remembered for long. But if it is highly emotional, it will be remembered, often for a lifetime. As also emphasized earlier and in Chapter 2, a complicating force is repression and activation of defense mechanisms to protect a person from negative feelings about self. Self is often the fulcrum around which emotional experiences are felt. Negative emotions may be repressed, probably by the hippocampus where cognitions and emotions associated with them are not allowed to reach full consciousness. Defense mechanisms outlined in Table 3.5 determine just what emotions will be remembered and the way a memory is experienced (through activation of body systems generating emotions) when recalled. However, repression of emotions often acts like a pressure cooker, with the repressed emotions increasing in intensity and exerting pressure to break out into consciousness. Yet, the emotions that emerge are often *transmuted* in a way to protect an identity or identities, which represents another cognitive mechanism for protecting self from the full effect of negative emotions. For example, a common pattern is the repression of shame experiences about self that, as they build pressure, come out as diffuse anger at objects that had little to do with the initial shame (through repression, plus displacement, as defense mechanisms). The same can be true of intense guilt, with anxiety and fear often being the emotions experienced but not in the original context where the guilt was first experienced. Typically, the anger is directed at objects that are safe in that they were not part of the original experience that led to repression and, equally important, at objects that cannot fight back. For example, spousal abuse is often the result of repressed shame and humiliation transmuted into its anger component, with anger targeting a safer object, such as a spouse who may not be in a position to fight back (with the original source of the humiliation, such as a boy's abusive father, getting a free ride while others must endure constant fear of arousing anger).

Thus, memories of past experiences with others and situations can become highly distorted by repression and then activation of a defense mechanism. The result is that the original experience will be highly distorted or not even be remembered. The emotions associated with the experience will not be the ones consciously felt by a person; original emotions may be activated but not allowed to escape the censors of the hippocampus as they interact with the prefrontal cortex. Thus, the hippocampus operates as a kind of sorting mechanism by accepting for longer-term storage only those memories that have emotional significance and that have been periodically recalled, thus activating the emotions tagging the cognitions holding the memory. This is normal and useful because it eliminates the need to remember unimportant or insignificant. We never remember, for example, all the times that we have filled our cars with gas, but if we shot gas all over ourselves on the way to an important occasion, the emotions aroused during such a routine activity will likely be remembered, at least in the medium term.

Repression of emotionally disturbing memories in reference to self-feelings is also a sorting mechanism, but it is less benign, and in fact, often burdens an individual with not only expending constant energy to keep the negative feelings from reaching consciousness but also managing the transmuted negative emotions that emerge as "emotional steam" from the hippocampus pressure cooker. Emotions such as diffuse anxiety and fear, anger, and sadness are indeed stressful and can cause health problems in the long run. This defensive regime orchestrated by the brain is, however, *natural to humans* with large brains, both neocortically and subcortically, that can hold long-term memories over a lifetime. Protecting self, which is the center of persons' relations with others and groupings, appears to have evolved with a brain that has the capacity to force individuals to live with unpleasant emotions for long periods of time. Thus, memories are ordered initially by the hippocampus but can also be repressed by the hippocampus if necessary for short-term equilibrium, thereby setting into motion complicated emotional dynamics.

Ordering Memories and Creating Stocks of Knowledge at Hand

One of the interesting things about the human brain is that what Alfred Schutz (1932) termed *stocks of knowledge at hand* are often difficult to recall consciously and recite or outline in any detail during talk, and yet, when situations require the memory, we appear to almost automatically invoke this knowledge. Even without direct experiences we somehow have acquired a necessary knowledgeability for a situation. I still recall my young son, who had never met his great-grandmother, attending her funeral; I was worried how he would act. I am not sure he had ever even observed or vicariously experienced a funeral in any way. And yet, when he arrived, he behaved in the properly solemn manner, shook hands (which I had never seen him do) with relatives whom he had never met, consoled them in an almost adult manner, began to cry when he came face to face with my grandmother laid out in the casket, and otherwise acted like a five-year-old adult. Where did he get this knowledgeability? I even asked him, and he looked at me strangely, implying that "anyone knows how to behave at a funeral."

In complex societies, being able to absorb and invoke such implicit knowledgeability is useful because of the many diverse situations where such knowledge may be needed. Still, this capacity was also essential in simple societies where the young were not so much taught anything but simply absorbed knowledge by being around adults and peers. Thus, it may be in our nature to seek information, almost automatically and even unconsciously, that our large neocortex somehow orders by potential salience for future use. Indeed, it may be the emotions attached to others as they reveal information to the young that activates the brain to absorb the emotions so that the cognition will take on greater salience.

Future/Potential Salience as an Ordering Mechanism

It is possible that humans, so reliant on their interpersonal abilities to construct lines of proper conduct with others, have a capacity to recognize interpersonal practices, such as proper demeanor at a funeral, as implicitly relevant to self *in the future*, with even tangential exposure leading them to remember what is interpersonally involved. It would be the emotions surrounding self that would tag the memory, or perhaps a sympathetic emotional experience could be invoked through role-taking and empathy if an event had been experienced in some way. By five years old, children see themselves as a certain kind of object, experience a wide range of emotions, and remember syndromes or strips of interpersonal demeanor that look to be highly salient, now and in the future. Thus, a lifetime of such experiences, even if only vicarious, orders stocks of knowledge for potential use if an interpersonal situation demands recalling that knowledge. The person may not be able to articulate this knowledge, except in a general and vague way, but when the situation calls for a need to protect and verify oneself as worthy, the emotions surrounding this need are perhaps enough to pull the appropriate demeanors out of the hippocampus and/or frontal lobe.

Abstraction and Response Generalization as Ordering Mechanisms

The large brains of humans allow for *abstraction* and *response generalization*. Both represent mechanisms by which the brain absorbs experiences and then is able to make them potentially relevant to a wider range of occasions. For example, solemn behavior at a funeral is, in essence, a syndrome of expressive behaviors that communicate the feeling of variants of sadness as the fundamental emotional state, requiring of a person expressive control of face and body demeanor in interpersonal behavior and in talk with others and often ritualized statements and body language communicating the serious and sad nature of the occasion. Parts of this kind of "funeral behavior" where *bereavement* is to be the dominant expressive display can be adapted to any situation where seriousness and tempered emotions are supposed to dominate interpersonal discourse and behavior. Thus, if syndromes of interpersonal behavior useful in generic classes of encounters with others can be abstracted and generalized, it is easier to store them in memories, along with the cues for which variants of this more general syndrome are appropriate in a given situation.

Similar forms of abstraction apply to interpersonal actions where happiness and its variants are to be displayed and can be generalized and abstracted in mental storage, with additional triggers for *what kind* of happiness is to be displayed (e.g., the kind of happiness to be displayed at a collective carnival, wedding, high-octane party, concert). Thus, many "stocks of knowledge at hand," to use Schutz's words, may remain implicit because they are stored as a *general syndrome* of emotionally charged behaviors, with specific markers and triggers for the variants to be invoked in a given situation. Again, it is the emotions that are to be felt and expressed that do much of the cataloguing of the general behavioral syndrome to be invoked and then adjusted to the situation. Thus, it is in the nature of humans to store stocks of highly generalized and emotionally ordered behaviors and, then, to apply them to types of interpersonal occasions while making fine-tuned adjustments to the particulars of the occasion.

Framing as an Ordering Mechanism

At about the time that Erving Goffman (1974) was developing a conception of frames and framing during interaction as a key phenomenological process, computer science was also developing a related idea whereby information would be stored and retrieved in terms of frames that delimit the range of information searched for. Goffman's vision of frames and framing was, I think, more complicated than it had to be and was probably too complicated to achieve what frames can do: denote what is to be part of the interaction and what is outside the "frame," allowing for a delimited focus on a narrow range of phenomena. Individuals initially, to use Goffman's term, "key" a frame to start the interaction. If frames are to be changed, then there are interpersonal practices for "rekeying" the frame that is to guide an interaction.

Keying and rekeying often appear highly ritualized in talk and body language, because rituals call attention to shifts in the interpersonal flow and activate emotions if these shifts are not honored by others (Turner 2002), which will be examined in Chapter 10. Moreover, Figure 10.1 on page 227 offers one way to conceptualize basic frames that can be keyed during interaction.

"Chunking" Information as an Ordering Mechanism

Framing and other interpersonal dynamics may be greatly influenced by the propensity of humans to order information in "chucks" and subcategories. In a world where information can only be remembered (i.e., a world without writing and storage systems from the cloud to libraries), humans remember more complex ideas by creating general categories that then contain subcategories, such that when the general categories are remembered, the information in subcategories can also be recalled. Humans will generally have trouble remembering long lists of almost anything (unless they have trained their mind to do so and, even then, they begin to "chunk" related ideas together so as to remember a whole list). There are also limitations on how many basic categories can be kept in play at any given time, apparently about seven. Thus, individuals typically will be able to store considerable information in seven basic "chunks" with each major category or "chunk" of information holding about the same number of subcategories that can be invoked once the major category is mentally brought into play.

In simple societies, even those developing an elaborate folklore, this level of "chunking" was probably quite sufficient, and individuals who could train their minds to remember large amounts of information could acquire extra prestige because of their special powers of storytelling, recalling folklore, or repeating creation myths, and other activities involving holding larger amounts of cultural information. With writing, formal education, and other learned procedures for storing information, coupled with the capacity to store most information outside of the neocortex, humans can retain a great deal of knowledge and then rely on external sources for storing that knowledge. With the advent of cyberspace and search programs such as Google, it is possible to rapidly retrieve an enormous amount of needed information from an incredibly large number of "chunked" categories. Still, most of what people need to conduct themselves in interpersonal situations, even in complex and fluid situations, is stored as "stocks of knowledge-ability" in the human brain. Indeed, in the midst of a fluid interaction, a Google search is not likely to help much because the information would not be ordered in the way needed for the rapid-fire adjustments that people must make in interpersonal situations.

Related to chunking of information is simple *categorization* as a process for ordering and storing of information. Cognitive categories distinguish key elements of information by placing related bits of information in one category while placing other bits of information in other categories. For example, as will be outlined in Chapter 10, humans categorize all interactions by the basic type, with three types being universal: (1) work-practical, (2) social, and (3) ceremonial interactions. Humans also categorize the nature of others into four basic types: (1) *intimate*, (2) *personages* whereby limited personal knowledge is known, (3) *categories* of persons, where all that is known is that the individual is a member of a categoric unit, and (4) *strangers*, where little or nothing is known about an individual. Each type of individual requires different greeting rituals, modes of talk, body language, positional juxtaposing of bodies in space, and other normative expectations. Just across these three categories of situations and four categories of type of person, a great deal of information can be stored and retrieved (note: the total is the magic number seven). For example, if the situation is work-practical and the interaction is with a personage who is also in a category of high authority, most individuals would have no trouble invoking the relevant stocks of knowledge and interacting with the person. Even when, say, at a social situation like an office party, where norms of informality and sociality apply, most people could draw down the relevant stocks of knowledge of how to interact with a figure in high authority in an informal, social occasion within a workplace. A lot of information can, therefore, be downloaded almost instantaneously, once again a testimony to what a large neocortex and prefrontal cortex powered by emotions from the subcortex can do.

Gestalt Dynamics as Ordering Mechanisms

Gestalt psychology is, of course, an old approach, as is its companion field theory. These perspectives outlined a number of mechanisms by which the brain orders cognitions (e.g., Lewin 1947, 1951; Heider 1946, 1958; Festinger 1957; Newcombe 1953). There are now vast empirical literatures on the dynamics outlined in these schools of thought, indicating that the gestalt intellectual movement was onto something important. Here we examine four that are particularly important in understanding human nature.

Cognitive Congruence and Consistency

Humans are motivated to seek consistency and congruence among cognitions because the lack of consistency and congruence (e.g., cognitive dissonance) appears to arouse negative emotions. Motivated to avoid negative emotions, humans are willing to adjust cognitions to achieve consistency and congruence. When cognitions are consistent, they are probably easier and more efficiently stored in memory. In fact, the level of consistency and congruence, or the lack thereof, may be one basis of ordering all cognitions. Those that are inconsistent may be stored in different locations because they are tagged with more negative emotions. If the cognitions are too inconsistent (e.g., "I am supposed to love my husband, but I actually hate him") and this inconsistency generates guilt and other negative emotions toward self, then these incongruent emotional states may lead to repression where the negative emotions are not felt but come out in a transmuted form as episodes of anger and high anxiety and perhaps extreme sadness. Experiencing negative emotions that arise from the repressed incongruence between love and hate can generate a new round of inconsistent cognitions that arouse a new round of negative emotions. These transmuted emotions that stem from repressed emotional inconsistencies often dominate minded deliberations on how to consummate impulses and make making them difficult if not impossible to consummate. The impulses then increase in intensity and dominate the perception and manipulation phases of Mead's "act," sending persons into a Freudian search for ways to repress the inconsistencies and protect self and particular social relations.

Contrast-Conceptions

Humans always notice differences. Numerous empirical studies document individuals' ability to notice and remember differences, suggesting its importance in the way the brain stores memories in terms of contrasts. The brain will thus remember (a) categoric units, such as black-white or male-female, (b) physical differences like tall-short, (c) status differences such as prestigious-stigmatized, powerful-weak, super-subordinates, (d) different situations as in work versus play, and (e) moral positions such good-bad or religious-atheist, politics like liberal-conservative, and many other contrasting properties of the social, biological, and physical world. When humans interact, the existence of contrasts and differences will almost always be highly salient, prompting individuals to adjust their demeanor in order to get through the interaction. For the most part, interactions are more dominated by contrasts that signal differences in categoric units (male-female, ethnicity, high vs. low social class) and differences in status within a corporate unit (high power–low power) because these contrasts are the most salient for sustaining the interaction (Turner 2002, 2007, 2010). Indeed, much knowledgeability is probably stored in this fashion as a series of varying expectations for how people should behave with individuals on one or the other side of a contrast. Humans are motivated to get along if they can, unless they have been mobilized emotionally to engage in conflict (Collins 2008). They will thus download the relevant stocks of knowledge that enable them to interact with differences, if they must and cannot avoid the interaction.

Expectation States

Humans tend to record outcomes of interactions. If outcomes are similar over time, humans develop expectations for outcomes into the future (Berger 1958, 1988; Berger et al. 1977; Ridgeway 1986, 2001). Indeed, the interaction complex holds that continuity in interactions over time relies on the formation of expectation states for how different people will behave, especially if status among them is different. Moreover, members of categoric units, such as males and females, will be subject to expectations often codified in cultural beliefs about how males and females should behave (Ridgeway 1986, 2001; Ridgeway and Erickson 2000).

Violation of these expectation states leads to negative emotional arousal, and because people generally seek to avoid negative emotions, expectation states have great power, even when they support unfair inequalities. To violate expectation states generally takes considerable emotional mobilization, often around anger over injustice and inequality, and once done, the encounter will be dominated by negative emotions.

Thus, a great deal of cultural knowledge affecting norms and beliefs is codified as expectation states, and these are often highly moralized. Individuals tend to be particularly angry when norms are violated or when individuals are subject to unfair stereotypes and expectations that demean them. Individuals may seek to challenge what is seen as immoral. The power of norms and beliefs resides in their moralization, which, if operating effectively, makes individuals feel shame and/or guilt when they violate them; therefore, it takes considerable anger for a person to embark on a personal crusade (often part of a more general social movement) to violate expectation states enshrined in cultural norms and beliefs. Still, without some expectation states, interactions will be stressful because individuals will not be able to relax in their role-taking. Humans store cultural codes as a series of expectations of self and others that, if accepted by all, dramatically reduce what has to be remembered and what interpersonal practices need to be invoked. Because expectation states are particularly likely to pertain to status differences in categoric and corporate units, they are critical to sustaining social structures and their cultures—thus making individuals even more likely to remember and invoke these expectation states during an interaction if they can do so without violating their personal morality.

Attributions

Humans almost always make attributions for their experiences, particularly those that have aroused emotions (Heider 1946, 1958; Piaget 1948, 1952; Weiner 1986). As noted in Table 3.5 on page 79, they will generally make self-attributions for positive emotional outcomes, whereas defense mechanisms may push them to make external attributions to others, situations, culture, social structure, or even unobservable objects like gods and supernatural forces to "explain" why they have experienced negative emotions (Lawler et al. 2009; Turner 2002). Attributions thus have *proximal* and *distal biases*. The more positive the emotions experienced, the more proximal will the attributions be, targeting self and perhaps others in local situations. Conversely, the more negative the emotions, the more likely the attributions will be distal and target "safe" external objects that cannot fight back against negative feelings. Thus, categories of others, corporate units, and more remote social structures will be blamed (Turner 2000, 2002, 2007; Lawler 2001). If, however, individuals make self-attributions for negative emotional experiences, then they will experience shame and guilt, as well as other negative emotions, thereby potentially setting off activation of defense mechanisms such as displacement and projection that protect the self.

Attributions are probably manifestations of a more fundamental cognitive propensity to reckon causality that may be a by-product of a large brain able to store memories and experiences. Indeed, attributions may be yet another sorting mechanism by which cognitions are remembered because they almost always are tagged with emotions, and especially emotions revolving around self. External attributions are, as noted in Chapter 3, more than simply a cognitive propensity; they are also a defense mechanism whereby individuals protect self by blaming more distal objects. Thus, once a cognitive process is so entangled with evaluations of self, it becomes fundamental to humans' emotional life and, thus, is an important part of human nature.

Conclusion

The literatures in sociology, anthropology, and psychology on these and other cognitive dynamics are extensive. My portrayal is delimited by my concern with human nature, especially as it affects interpersonal processes as these build up, sustain, or change social structures and culture. Moreover, because the larger human brain is, in its origins, tied to prior increases in emotional capacities of humans for meeting selection pressures for forming strong social bonds and group solidarities, my concern is also with those cognitive processes implicated in emotional arousal of individuals during interaction.

To function properly, a large brain needs to order larger amounts of information. This process begins by tagging emotions to remember experiences and then, by virtue of emotional tags, retrieving and using the experiential information in decision making—all of which promote fitness. There are limits to how much information can be stored, especially information not directly related to interpersonal behaviors. The ordering mechanisms discussed in this chapter are, I believe, part of humans' biological nature because they evolved as emotions were allowing for growth in the neocortex, development of speech, and the use of speech to create culture. In preliterate populations, this knowledge and culture had to be stored in the frontal lobe and retrieved by the prefrontal cortex. At the same time, in simple preliterate societies, the full suite of inter-personal capacities of humans was operative. Members of such societies were not burdened with interaction in complex and differentiated social structures with more robust cultures. The neurological capacities of humans to create and sustain rather large and sophisticated "stocks of knowledge at hand" for interaction in large, complex societies was only possible because early humans possessed the neurological wiring to order information along the ways that outlined in this chapter. Even more cognitive dynamics may be involved, but certainly those reviewed here are part of human nature because modern, complex, and ever-changing societies like those in the present era could never have evolved without these cognitive capacities fully in place.

The biology-based properties of human nature evolved under one paramount selection pressure: get better organized by forming strong social bonds and group solidarities. What emerged was bigger sub-cortical emotion centers allowing for a larger neocortex that, in turn, led to full utilization of humans' capacities for language inherited from LCAs of hominins and ancestors of present-day great apes, and then the accumulation of culture. By going back in time to look at the preadaptations and behavioral capacities/propensities of LCAs (Chapter 2) as well as by comparing key brain structures of great apes and humans (Chapter 3), we can get a real sense for what humans inherited from their primate ancestors. Then, by examining the su-percharging effects of enhanced emotions, larger brains, spoken lan-guage, and culture on these inherited traits, we have been able to isolate as best as is possible the biological basis of human nature and, to a degree, separate it from the effects of sociocultural formations on human behaviors and cognitions. Emphasizing emotions and interac-tion as the basic force allowing humans to organize focuses on human nature *sociologically* rather than psychologically or in the narrow sense of biology. It may have seemed strange to title a book *On Human Nature* with the subtitle: *The Biology and Sociology of What Made Us Human*, when so much has been about the biological basis of human nature. Here, biology and neurology are examined from a sociological perspective that views the fundamental nature of humans as shaped by *pressures for the ancestors of humans to get better organized,* or die. This line of emphasis is typical of neither psychological nor biological analyses of human nature and its evolution.

Figure 7.2 fills in the box at the right of Figure 6.2a on page 130 for the "evolved cognitive complex" with a more abstract listing of the cognitive complex described in this chapter. Human nature is significantly affected by the dramatically expanded cognitive capacities of humans, compared to great apes and certainly most, if not all, animals on earth. Our world is mediated by emotionally tagged cognitions stored in complex ways, retrieved, and then assembled rather rapidly. Moreover, a large neocortex and speech allow for a world mediated by culture, with virtually all aspects of our experiences ordered and often regulated by cultural proscriptions and prescriptions and by shared meanings built up over time and transmitted across generations.

The mechanisms for ordering information cognitively should be seen as part of human nature because they help us understand not only how cognitions are stored and perhaps retrieved but also how cognitive mechanisms have large effects on all the other complexes that make up human nature: emotion, interaction, psychology, and community complexes. True, none of this cognitive complexity would have been possible without prior changes in the subcortex to make hominins and then humans more emotional, but once these cognitive dynamics are in place, they have large effects on emotional arousal, on interaction, on fundamental psychological need-states, and on how the orientation to community as the basis of social organization has been extended during societal evolution.

The reverse causal arrow in Figure 7.2 from the evolved cognitive complex back to the elements of the elaboration machine represents an important reverse causal effect. As the neocortex evolved, it not only allowed for speech and culture to become cognitive capacities; it also fed back and increased the dynamism of the emerging elaboration machine by enhancing emotions that allowed for further growth in the neocortex for speech to evolve, which, in turn, led to evolution of culture. And so, each of the elements inside the elaboration machine not only fed forward during hominin and early human biological evolution, they also revealed reverse causal effects, with culture increasing emotionality, cognitive growth, and capacities to use language to enhance culture. Similarly, speech per se increased capacities for thinking, deciding, and remembering, while also increasing the ability to denote new emotions. Thus, the list of elements in the cognitive complex are not static; they are a part of a dynamic set of complexes affecting each other and the operation of the elaboration machine. This is the reason that the brain, especially the neocortex, suddenly began to grow so rapidly 1.0 mya and perhaps as late as 0.7 mya, leading to the emergence of humans about 450,000 years ago. Once elements of the elaboration machine were falling into place, their effects on each other accelerated biological evolution, which, in the end, made possible the evolution of human societies while at the same time altering and enhancing the inherited biology from the LCAs of hominins.

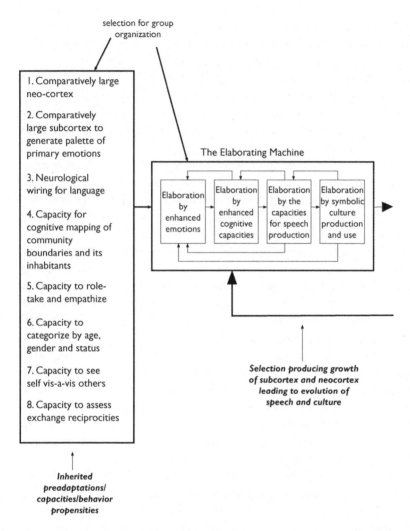

Figure 7.2 The Evolved Cognitive Complex and Human Nature

1. Large neocortex, fueled by subcortex, creates the capacity to store and order information and experiences tagged with emotions into large stocks of knowledge at hand, thereby making information available for retrieval by the prefrontal lobe.

2. Language and symbolization of all experiences, accelerate and expand capacities and allow for further creation, accumulation, use, and transmission of shared cultural meanings as memories and stocks of knowledge that increase fitness and adaptation to both the bioecological and sociocultural environments.

3. Formation and ordering of stocks of knowledge through cognitive capacities for abstraction, attributions (as to sources, causes, and origins) of events, salience of identities, consistency and congruence among cognitions and emotions, contrast-conceptions, cultural coding of prescriptions and proscriptions, cognitive partitioning of inconsistent information and emotions, expectation states, and activation of defense mechanism.

4. Formation of moral cultural codes through speech, while the capacity for identity formation leads to moral evaluations of self and self-sanctioning by activating shame and guilt as emotions of social control.

5. Prefrontal cortex in interaction with hippocampus and memories stored in the frontal lobe allows for repression of cognitions tagged with emotions directed toward self and for activation of defense mechanisms—displacement, projection, reaction formation, sublimation, and attribution—that protect self and transmute (a) the nature of the emotions experienced and (b) the targets of these emotions.

6. Capacities to read the gestures of others to determine their dispositions and likely courses of action during an interaction while at the same time using speech and nonverbal gestures to communicate internal states and likely courses of action, in a process of "taking" account of

others' internal states and likely paths of action and, simultaneously, presenting self in ways "making" others
aware of internal states and likely courses of action in a
process of mutual (a) role-taking and making, (b) status-
taking and making, (c) identity-taking and making, (d)
emotion-taking and making, (e) culture-taking and making,
(f) structure-taking and making, and (g) situation-taking
and making.

7. Retrieval of emotionally tagged cognitions in order
to mutually (a) categorize others, self, situation, cultural
expectations, structural constraints, and situational expectations, (b) use appropriate speech forms and expressive gestures in opening, forming, and closing the flow
of interaction, (c) invoke appropriate keys and rekey the
frames during interaction, (d) assess which need-states of
self and other can and/or should be met or not met in the
situation, and (e) assess resources to be exchanged and
invoke norms of fair exchange.

Note

1 Anthony Demasio has proposed, not without some criticism, the *somatic marker hypothesis* that when experiences are remembered, the emotions associated with these cognitions are also remembered, and moreover, are activated in ways that mobilize the body systems—neurotransmitter systems, neuro-active peptide and more general hormonal systems, muscle-skeletal system, and autonomic nervous system—that were activated during the episode of interaction when the memory was originally laid down, thereby re-creating in a somewhat diluted form the emotions associated with a memory or experience. It is this remobilization of the body systems that make memories more likely to be remembered again, and if remembered, eventually passed from the hippocampus to the frontal lobe for longer-term memory storage.

8

The Evolved Emotions Complex and Human Nature

The capacity and propensity for humans to be so emotional are as distinctive and unique as our cognitive capacities and, indeed, are the reason that the neocortex grew and made humans not only the most emotional animal on earth but also the most intelligent. Using the inherited preadaptations and behavioral capacities and propensities of the last common ancestors (LCAs) of hominins and present-day great apes, natural selection began to increase key nuclei in subcortical structures of the brain to enhance hominin emotionality and, thereby, increase social bonds and group solidarities. In so doing, selection installed the engine of the "elaboration machine," which allowed for a larger neocortex and greater intelligence to be fitness enhancing, thus initiating construction of the second element of the elaboration machine. With the slow building out of the emotional and cognitive components of the elaboration machine, speech and then symbolic culture could operate much like turbocharges that accelerated hominin evolution leading to the first humans—*Neanderthals*, *Denisovans*, and *Homo sapiens*—that would provide the genome for humans today. Thus, the elaboration of emotions, as outlined in Chapter 3, began the construction of the elaboration machine, leading to a larger subcortex and then neocortex, an auditory or speech-based language system, and a system of cultural representations of ever more dimensions of the universe. Once the first build-out of the elaboration machine existed by late hominins, such as *Homo erectus* or *ergaster* (2.2 to 0.3 million years ago [mya]), the machine became a perpetual motion machine, pushing for further neurological development of emerging humans. Later, it pushed for further development and elaboration of the sociocultural formations made possible by the dynamic elaboration of cognitive, linguistic, and cultural forces that could have feedback or reverse causal effects on the original emotional elaborations that had started the construction of the elaboration machine. Humans' emotional capacities could thus become further elaborated and, in so being, have additional effects on cognitions, speech, and culture.

Brain Growth, Language and Speech, Culture, and the Elaboration of Emotions

The expansion of humans' emotional capacities—that is, the number, variety, valancing, nuance, and intensity of emotions—made growth of the neocortex fitness enhancing because more cognitions could be held in memory, retrieved from memory, and used in decision making, thereby promoting fitness. Having a larger brain that can tag more complex cognitions with more diverse and varied emotions made humans smarter. This relation between expanded emotions and larger brain functioning probably took several million years to evolve along the hominin line—from *Australopithecines* some 4.0 mya to late *Homo erectus/ergaster*. Most of the brain growth leading up to the level of the human measure occurred over the last million years of hominin evolution, moving rapidly (in evolutionary time) to increase overall brain size of late hominins and early humans from around 500 cc to 1,350 cc on average for *Homo sapiens* and even larger for *Neanderthals* and *Denisovans*, even as the brains of some more isolated demes of various species of hominins' brains had not grown significantly beyond 500 cc—as was the case with *Homo naledi* and some subpopulations of *Homo erectus*.

The brain does not think in language as we know it. If it did, we would think very slowly since speech is sequential, with phonemes and mor- phemes ordered by grammars over time. Human thinking is dramatically more rapid, simultaneous, and gestalt based than speech. For this reason, humans have *Wernicke's area* to upload sensory inputs (including auditory speech sounds) into the brain for rapid processes in "the brain's way of thinking," and reciprocally, why *Broca's area* for downloading brain thinking into sequential speech evolved (see Figure 3.1 on page 60). Speech is what allows for human culture to evolve because any aspect of the environment, any thought or emotion, can be labelled and integrated with other thoughts and then shared with others through speech acts. Speech is also both denotative and connotative, which allows humans to label and point to dimensions of their external environments as well as to their inner environment of thought and, then, to draw out further implications of what is denoted. For example, to tag an emotion with the label *love* em- phasizes a particular variant and property of happiness, while at the same time, speech acts can elaborate on the emotion of love to produce new variants—for example, what love means, what kinds of objects can receive love, what is not true love, what is problematic love, what is dependence- based love, when is love appropriate or inappropriate, what is passionate love, and how is love conflated with other emotional states like the pleasures of sex, friendship, parenthood, and so on. Moreover, speech can lead to the development of *emotion ideologies*, such as what "true love" is supposed to be in a given culture. It can even stimulate cultural social movements as occurred in the West with the development of theater and novels devoted to expanding on the range of emotions associated with love.

When talk can be written down and systematized into moral codes and folklore in the culture of a population, talk and speech elaborate the variants of emotions available to humans. In other words, talk and speech extend emotions and connect them in ways that may not have a specific neurological basis but rather a cultural basis, tied to cultural beliefs, values, ideologies, and norms inhering in culture rather than human neurology.

For a long time, a debate centered around whether emotions were cultural constructions or neurology-based states. Sometimes, authors argue that the primary emotions have a neurological basis, but their elaborations (see Tables 3.2, 3.3, and 3.4 on pages 62, 66, and 76) are cultural, whereas others argue that most emotional states are physiological and tied to activation of neurological structures in the brain. Most sociologists probably come down on the constructivist side of this argument, although I tend to come down on the more physiological side of this long-running debate. (For neurologists who are moved to the constructivist side, see Brothers 1997.) Either way, a larger brain using language and culture can expand and elaborate emotions beyond what body systems—that is, neurotransmitters, neuroactive peptides, hormones, autonomic nervous systems, and musculoskeletal system—would normally do alone without speech and culture. In fact, culturally designated emotions or even emotions aroused by individuals in speech acts will activate these same body systems generating emotions. Thus, in this limited sense, social constructions of emotions have a biological basis, but the activations of these emotions come from elements of the elaboration machine. As emotions become reflexive, or subject to reflection and discussion, new variants of emotions are created, and these can feed back and activate neurological systems that activate the body systems generating emotions. Thus, even emotions that have been culturally constructed through talk, discourse, and reflection *must* activate the neurology of humans for these new emotions to have any real power over individuals or groups of individuals. Moreover, speech also allows for the construction of emotions unique to individuals, as is the case when a person "talks" to herself or himself about feeling hurt, mad, sad, and perhaps also relieved about a love affair gone wrong. Here the person is, to a degree, activating the body systems producing these emotional states and, in essence, creating perhaps a unique or idiosyncratic emotional state that may become part of a person's unique palette of emotions as well as structure a person's memories stored in the hippocampus and frontal lobe. If this new second-order emotion is passed on to others in further talk or writing, then it may become a part of subculture in a population.

These emotions are elaborations on the base laid down by natural selection in rewiring the hominin and then human brain for expanded variants of primary emotions, for new first-order combinations or elaborations among two primary emotions, or second-order elaborations of three primary emotions, such as the examples outlined in Tables 3.2, 3.3, and 3.4 on pages 62, 66, and 76, respectively. Expanded emotions allowed for brain growth and what such growth would facilitate: the evolution of spoken language and culture. With larger brains, spoken language, and culture, new elaborations of emotions can be experienced and even expressed among individuals. These emotional and cognitive propensities for elaborations *are*, I believe, part of human nature because they are made possible by biological systems in the subcortical and neocortical areas of humans' brains. A large brain is part of human nature; and its evolution has expanded humans' capacity to experience and express their world with not just cognitions but also *cognitions about emotions and, conversely, with emotions about cognitions* that feed back and forth into the neurological systems, activating the body systems that produce the sensation of emotions and the systems of ordering, sorting, and storing cognitions (see Jackson et al. 2019 on the effects of culture and speech on the biology of emotions).

There probably is an upper limit to how many emotions can be elaborated because too many emotions would slow down thinking. The range, nuance, and varieties of emotions that humans experience can be expanded, however, which is part of human nature because it has a biological basis: a brain programmed to use language, a brain with dedicated nuclei (Wernicke's and Broca's areas) that can upload and download speech into the brain's mode of processing information, and an evolved set of neurological structures (along the Sylvian fissure and, to some extent, the cerebellum) regulating the physiology of speech production (larynx, tongue, lips, and associated muscles).

Thus, driving much of hominin evolution under increasing pressures to further social ties and group solidarities, natural selection made humans the most emotional animals on earth. As mentioned at the outset of this chapter, humans' emotional capacities are as unique as their cognitive, cultural, and linguistic abilities. If emotions were necessary for brain growth, language, and culture, then reciprocally, larger brains, language, and culture enabled humans to become even more emotional in ways that support, embellish, and give power to cultural values, morals, ideologies, norms, and stocks of knowledge. Such elaborations of emotions feed back into the neurological systems generating emotions in the first place. This elaboration and loop from emotions to culture and back *is what it means to be human* and hence is part of our fundamental nature.

Emotions and Reflexivity

The elements of the cognitive complex ensure that humans are reflexive and can think about themselves in past, present, or even future contexts. If they want to slow down the process of reflexivity, they can talk to themselves. Indeed, the process of downloading thinking about self or any topic into sequences of speech is often one way to dramatize and highlight cognitions and emotions. Emotions will force reflexivity because when humans think about themselves, their experiences, their relationships, their biography, or their life in general, the arousal and flow of emotions almost always leads to further thought (and often more emotional arousal). Thus, the more cognitions about any dimension of human experiences are tagged with emotions, and the more intense these emotions, the more individuals will think about experiences and what they mean in the past, now, or potentially in the future. This part of human nature results from larger brains that can only hold cognitions that are valanced with emotions. The capacity for self-talk often leads to the elaboration of new emotions or shifting valences on existing emotions, ensuring that humans will think about themselves in terms of the new emotions, often aroused by self-talk (Jackson et al. 2019).

Reflexivity about situations in group contexts will also lead to the formation of cultural codes to regulate expected behaviors. Interpersonal capacities of humans allow for mutual reading of emotions, along with speech, as individuals role-take and status-take with each other (that is, discern each other's dispositions to act in certain ways and their status vis-à-vis self). As interactions are regularized, they become *expectation states* that, if invoked repeatedly, will generally be enshrined in normative codes, which become one more emotionally tagged set of cognitions stored collectively among individuals. In this way, they become part of the culture of a population or cultures of subpopulations. When normative expectations are realized, individuals experience low-level satisfaction, or even more powerful positive emotions in the face of concern about whether norms would be followed (Turner 2007). When violations of expectations occur, however, negative emotions are aroused—variations of anger and perhaps fear as well—while those failing to meet expectations may experience variants of shame and guilt, potentially transmuted by repression and defense mechanisms into diffuse fear, anger, or sadness.

Arousal of negative emotions in encounters among humans inevitably leads to reflexivity, whereby individuals think about why others have violated normative expectations or where the person who has violated norms will engage in self-assessment and consideration of options of what, if anything, to do about the breach to the interpersonal flow. In contrast, when mild positive emotions ensue from successful interactions, less reflexivity is likely—unless there has also been expectations for problems in the interaction and, if such was the case, individuals will experience second-order elaborations of emotions like relief. Even an emotion like relief, which contains elements of the negative emotion of fear, can set off another round of reflexivity of why and how expectations were realized.

Thus, large brains, speech, culture, and emotions all work to make humans highly reflexive, oftentimes to the point of some immobilization as conflicting emotions and cognitions are sorted out. If such is the case, negative emotions such as worry, concern, fear, and anger can be experienced. Once again, the cognitive ability to think and ponder can generate negative emotions as individuals think about, and indeed, perhaps overthink about, events. At times such behaviors can be labelled *neurotic* or some similar negative statement that only arouses more negative emotions, but such is the price to be paid for a species whose nature is driven, in large part, by emotions attached to cognitions among big-brained animals that can think and talk to themselves about why certain emotions have been experienced. An animal with few emotional states that have not been elaborated by language would worry much less and not think as much about self and experiences. So it is in human nature to worry and think about what did or might go wrong in an interaction.

Emotions and Social Control

This centrality of emotions to human nature is also an underlying force of social control. When normative systems and other cultural codes are in place, they create *expectation states*, as do memories about what transpired in a past interaction with particular others. Whether as cultural prescriptions or proscriptions, or as memories of what last transpired, expectation states are generated not only in self but, equally important, in others. Thus, others expect certain behaviors and a failure to meet this expectations may signal negative emotional sanctions. Individuals who have been sanctioned will generally feel negative emotions, ranging from anger at, or disappointment in, the negative sanctioning by others. Conversely, if the person accepts the sanctions in light of the expectations violated, they will feel shame in not measuring up to others' expectations. In addition, if moral codes are also salient elements in the sanctions of others, they may well feel guilt as well.

Emotions themselves can become part of the expectation states on others. There are almost always implicit *emotion ideologies* or, at the very least, emotional normative expectations for how individuals in various types of situations should behave, especially in their expressive behaviors. To express the wrong emotions—for example, laugh at a funeral, joke during a religious ritual, criticize on a first date, express anger at a good friend, be sad at a celebration—is to violate more general norms and the norms of the *emotion culture* in play (Hochschild 1979). Along with any set of normative expectations comes specific emotional expectations about what emotions can and cannot be expressed, and when appropriate emotions are expressed, how they are to be expressed by self to others in a situation.

The emotions of guilt and shame, appearing rather late in hominin evolution (Turner 2000, 2007), evolved to solve a fundamental problem of great apes trying to get better organized in order to survive in more open-country habitats. This problem is that negative sanctions by others, for any reason, may also invite counter-anger from the person being sanctioned. Anger and counter-anger are disruptive to social relations and group solidarity. For an animals like hominins and humans who do not have powerful bioprogrammers for group formation that can override negative emotional arousal and even conflict, negative sanctions can reduce rather than increase the fitness of the group. Sometimes sanctions only generate sadness or fear that do not lead to conflict, but in both of these cases, such emotions are not conducive to forming strong bonds and group solidarities for the long run. Natural selection thus gave late hominins or perhaps only early humans something that present-day great apes do not experience: the arousal of guilt and shame. Guilt and shame do not promote strong bonds and solidarity, per se, but they do motivate individuals to avoid these painful emotions. In doing so, social control moves from external control (with negative sanctions from others) to internal control of the person who will sanction self. If guilt and shame are rarely experienced but feared, they operate as a powerful force of psychological control as emotions that attack self and identity unlike any other human emotions. They are powerful because they were probably needed, in many ways, as a substitute for the lack of biopro-grammers for group-oriented behaviors (in such structures as herds, pods, packs, troops, prides, etc.) and strong bioprogrammers for kin selection (where kinship ties always take priority). Guilt and shame are also generalized emotions that can be adopted to any social context or situation; they have flexibility that would always be fitness enhancing. Equally if not more important, they give sanctioning "teeth" to cultural proscriptions and prescriptions, which would be especially critical in sustaining social control in a culture-using animal like humans. Humans are, to some loose degree, controlled by their genome, but they are *not organized by strong drives emanating from their genome*. By creating a biology-based duo of emotions, combining in a second-order elaboration (see Table 3.4 on page 76), shame and guilt provide a powerful biological force in making culture-based social control viable. Guilt and shame are thus a significant part of humans' evolved nature because they give culture its power to control without external sanctioning, on the one side and on the other side by motivating individuals to avoid experi-encing these emotions and thus abide by expectations. Perhaps the reason that humans but not great apes experience shame and guilt is because such emotions are only possible when an animal can articulate cultural proscriptions and prescriptions that are then "enforced" by shame and guilt as much as external negative sanction from others. Moreover, only animals with a sense of self can experience guilt and shame because such emotions require self-awareness vis-à-vis others and situations requiring culturally specified behaviors of an animal with verification of identities on the line in each and every interaction with conspecifics.

Emotions and Self

Running through this discussion is the centrality of self as part of human nature. Other animals—such as dolphins, whales, great apes, elephants, pigs, and probably intelligent birds like some species of parrots and macaws—can see themselves as an object in their environment. Yet, for most species of animals that need to be group oriented, bioprogrammers are the most efficient way to sustain the group and species fitness, even if bioprogrammers do reduce flexibilities in responses to environmental change. For a species like *Homo sapiens* without strong bioprogrammers even for family groups, much less other types of groupings, some other kind of anchor or focal point for sustaining groups is required. The evolved emotionally charged self, mentioned in the cognitive complex and expanded on in discussion of the psychological complex in the next chapter, was critical to hominin evolution because, without the evolution of various emotionally valanced identities, there would be no "hook" to pull individuals into conformity with group-level expectations. It is in the nature of humans to be self-centered in at least this sense: we seek to have various identities verified, and this need-state is a powerful force of human nature. The only real way for this verification of self to occur is through interactions in group contexts where normative expectations and other cultural directives are operative and, at the same time, where others also seek to have their identities verified. Great apes are highly individualistic, even though they are quite adept at interpersonal relations when needed (thus giving rise to the interaction complex to be examined in Chapter 10). Thus, all efforts at social control—whether externally administered as sanctions by others or internally felt by the capacity of humans to have a conscience driven by the emotions of shame and guilt—focus on, and run through, self and various identities that individuals are trying to have verified in most situations. Without the rewarding power of positive emotions that arise from conformity to expectations or the punishing power of negative emotions, especially shame and guilt, from failures to live up to expectations, social control of individualistic evolved great apes would be problematic, especially when hominins were forced to live in ever more open-country habitats where social control and group stability were necessary.

The cognitive capacity to hold multiple selves or identities may have evolved as a by-product of neocortical growth, but it is the emotions tied to these identities, and especially emotions experienced through success or failure in getting identities verified by others, that make identity formation central to social control. For a species of mammals without strong bioprogrammers for group formation, the elaboration of self into a series of identities valanced with self-feeling and needs, along with cognitive elements specifying the nature of each identity, provides a powerful platform for social control. Natural selection not only expanded the cognitive capacity to hold multiple identities, it also generated, as part of human nature, emotionally charged cognitions about self. Moreover, selection also installed the capacity for humans to experience guilt and shame about self, thus making self the center of social control. In addition, natural selection generated powerful need-states for individuals to have their sense of self verified by others in order to experience positive emotions and to avoid shame and guilt. Thus, as part of the psychological complex, the most powerful motivating force among humans, perhaps save for motives for self-preservation and sex, is to have one's sense of self verified. Human self-preservation is more than life or death; it is also about having identities verified in a variety of contexts. Hence, the evolution of identities and motivation to have them verified was only possible by the valancing of identities with emotions and by the painful emotions experienced by self when self is not verified. For animals that can remember the past, negative experiences in verifying self or in experiencing chronic shame and guilt are highly pathological and often lead to the activation of defense mechanisms. This fear of losing self in this way motivates individuals to meet individual and group expectations enshrined in culture. Individuals learn the costs of failures to verify identities—memories filled with negative emotions that include sadness, regret, hurt, betrayal, anger, fear, shame, humiliation, and guilt. Some level of identity is on the line in most interactions, even when mediated as in social networks like Facebook where failing to click a "like" or, even worse, "defriending" a person can expose this vulnerability of all humans to identity verification. Whether in talk with others in actual groups or self-presentations in social media postings, *self is always on the line.* The nature of being human involves constantly experiencing some concern or *anxiety* about verification of identities. This quiet anxiety is also one of the most powerful mechanisms of social control that natural selection installed in the human genome, as perhaps only a by-product of making an animal that is emotional, big-brained, and self-aware. Even though large brains enhance fitness, they are stocked with emotionally charged cognitions, especially about self, that make all humans psychologically vulnerable.

Natural selection, whether by direct selection or as a by-product of making humans emotional and intelligent, also created the capacity for the brain to repress negative emotions and activate defense mechanisms listed in Table 3.5 on page 79. Defense mechanisms can situationally protect self, but if chronically activated, they make humans once again vulnerable to severe psychological pathologies. A big-brained, emotional, and self-centered animal like humans will always be vulnerable when failing to meet expectations, especially for verification of identities, because these dynamics are a basic part of human nature.

Emotions and Social Structures

The only permanent social structure among the ancestors of present-day great apes and humans' hominin ancestors was the community, or a home range that could be many square miles. Groups within this home range were always temporary, forming and then disbanding, with individuals often moving about alone only to hook up in temporary groupings. This kind of fluid, fusion-fission system of relationships was possible because of the interpersonal capacities of present-day great apes' ancestors, as is still evident in great apes today. Such a weak-tie structure was fitness enhancing in a more arboreal habitat where danger from predators is less and food sources are spread out, which require low population densities at any given point in the habitat.

As selection began to enhance emotions to create stronger ties and group solidarities among hominins living in more open-country habitats, the key bonding mechanism was the flow of positive emotions among conspecifics, eventually leading to more permanent groupings and the nuclear family that would be the structural backbone of the hunting and gathering band. As the other complexes of human nature evolved during this process of using emotions to forge groups, the growing neocortex of hominins would lead to speech and cultural production. Then, for perhaps as long as 400,000 years, early human societies consisted of bands of hunter-gatherers composed of varying numbers of nuclear families wandering a territory (the home range or community). Cognizance of the boundaries of the home range was probably a hardwired propensity of early hominins because it is clearly evident in great apes, particularly in humans' closest relative, chimpanzees. Emotional enhancement would likely increase the sense of attachment to a home range as well as increase local group solidarity and stability within the community. Yet, when individuals or families within a hunting and gathering band come into conflict, one or both families simply leave and migrate to another band—a propensity also made possible by the high interpersonal skills of the LCAs of hominins and present-day great apes. Moreover, when adults in nuclear families reveal conflict, all hunting and gathering populations have simple rules and rituals associated with separating adults, or in essence effecting a "divorce," with one or both parties migrating to another band.

The result of this use of emotions as the primary mechanism of group attachment, coupled with the interpersonal skills to be discussed in Chapter 10, resulted in a unique assortment of emotional, interpersonal, and cognitive capacities that would allow patterns of human social organization to grow, as has been the case over the past 12,000 years. No other species of mammals could create such large-scale structures because of bioprogrammers for group-level formations—packs, hordes, pods, herds, and so on—and for kin selection biasing group formations around related kindred. The most intelligent mammals on earth, such as dolphins and elephants, create larger communities and evidence a fusion-fission plan of group organization much like the great apes. Strong bioprogrammers governing structural organization of a species place limits on what selection can do if larger social structures become necessary in changing environments, whereas for early humans, there would be few such limitations. Even the programmed sense of community and mapping of this community was accompanied by weak ties within the community. The result is that the scale of human social organization could sporadically increase and, then over the a few thousand years, lead to the formation of mega societies of billions of inhabitants spread across large territories.

Emotions, cognitive orientations to the larger community, interpersonal flexibility, and *lack* of powerful kin selection or grouping bioprogrammers were the keys to this capacity to expand dramatically, when needed, the scale of human societies. Emotions, when directed at social structures and when reinforced by interaction rituals (see Chapter 10), are the force driving attachments and commitments to social structures and their cultures. Cognitively, humans were already oriented to larger territorial expanses by their cognitive ability to map the boundaries and demography of their community *as a whole*. Thus, without the normal bioprogrammers driving an animal to focus on more micro-level groups of most mammals, hominins and then humans were already programmed to *look beyond temporary groupings* to the more inclusive and expansive home range or community. If emotions are added to this mix, then hominins and humans could possess the capacity, if needed, to stretch their cognitive horizons and develop attachments to even larger sociocultural formations.

With the development of speech and culture, these larger sociocultural formations could be designated and cultural traditions and cultural prescriptions and proscriptions could evolve for each of these formations. In turn, emotions would allow individuals to form emotional attachments to many new layers of social structure, beyond even the community. It is no coincidence, then, that hunting and gathering societies became community based once the size of their populations began to increase beyond what bands of hunter-gatherers could organize. Communities are *the most natural structural unit* for a great ape, including an evolved great ape like humans, who, even today, often spend a great deal of time searching for "a sense of community" or bemoaning the "lack of community." As for humans, we can see what bigger brains, enhanced emotions, speech, and culture do to our orientations to community when they create often nostalgic emotional attachments to one's hometown or present community of residence. It is a natural attachment that humans do not even need to work hard at creating or sustaining, because *it is the only hardwired propensity to reckon social structure* that humans inherited from hominins. When enhanced emotions, expanded cognitive power, speech, and culture are added to this hardwired propensity to reckon community, however, attachments become more intense and the potential for their extending attachments to other layers of social structure increases. And so, the inherited capacity to look beyond the group sets up the potential for both cognitive awareness of, as well as mapping of, ever more remote social structures and new levels of emotionally charged attachments to these structures and their culture.

It is in humans' nature to reckon the social structures in which daily activities are carried out, from work, familial, educational, and social groupings lodged inside of communities or organizations. These structures, in turn, are attached to these various institutional systems (e.g., economy, kinship, religion, education) that are the building blocks of still larger societal and intersocietal systems. Each level of organization carries cultural traditions and regulatory codes, and individuals develop emotional attachments, or emotional aversions, to various levels of social organization. These are never neutral, and humans are always aware of their external presence. The emotions experienced at various levels of social structure influence how commitments to these structures will vary. Much depends on what occurs in interactions in the groupings at diverse locations in various structures and on the degree to which humans' fundamental needs are met (see next chapter on the *psychology complex*), thereby arousing either positively or negatively valanced emotions. As outlined earlier, experimental data by Lawler and associates (2001, 2009) indicate that when groups are organized around divisions of labor where individuals cannot assess contributions of each member's respective contribution to outcomes of the group, or choose not to do so in order to sustain positive emotional flows and solidarity, members will generally make attributions to the group rather than to persons, which, in turn, will increase commitment to the group. These commitments can travel from local group to organization and community to institution to society, and even to intersocietal systems. Humans' cognitive reckoning of social structures can thus move ever outward as larger sociocultural formations evolve beyond community. However, it is still the *emotions aroused in the more face-to-face interactions* at the group level that affect the intensity of commitments that they will make to macro structures (Collins 1975, 2004), which determine the extent to which larger-scale social structures will be viable in the long run. The first micro-level hunting and gathering societies were held together by a combination of bioprogrammers to look beyond the group; similarly, present-day macro or mega societies are ultimately held together by the same emotional dynamics generated in interaction, as we will see in Chapter 10 on the *interaction complex*. Social structures are, in the long run, only viable if they generate positive emotions during social interaction that allow individuals to meet fundamental human need-states (see next chapter) and, at the same time, generate positive emotional arousal directed at the group and beyond. Emile Durkheim (1912) came to this same conclusion in his later work on rituals directed at totems symbolizing societies, while more contemporary scholars such as Collins (1975, 2004) have filled in many of the details to this argument (see Chapter 10).

Human social structures are thus not like an ant or any insect colony where inhabitant behaviors are genetically programmed. Human micro and macro societies are *constantly in the process* of being *created and sustained* by the same forces operating on the individual and interpersonal level of social reality: positive emotional arousal directed at social structures and the cultures of these structures. Because social structures are conduits of emotional flows, they are vulnerable to the disintegrating effects of negative emotions. Thus, the viability of social structures and their cultures ultimately depends on the emotions aroused among humans at micro levels of social organization.

Conclusion: The Emotions Complex

The *emotions complex* was, I believe, the first of the complexes of human nature to elaborate beyond the inherited legacy from the great ape evolutionary clade. Once subcortical areas of the brain started growing, something like the elaboration process of emotions outlined in Tables 3.2, 3.3, 3.4, and 3.5 on pages 62, 66, 76, and 79 ensued. In turn, this enhancement of emotions kick-started neocortical growth that eventually led to the evolution of speech and symbolic culture. Once these other elements of the elaboration machine were in place, they had powerful direct and reverse causal effects on emotions, expanding them ever further as many new emotions were idiosyncratically constructed by individuals thinking and talking to themselves about their feelings; and when doing so more collectively, such speech acts eventually created emotion cultures attached to all layers of social structure in human societies. The content of most elements of human culture in general—including oral traditions, religious beliefs, norms, values, ideologies, technologies, and so on—vary from context to context, but one constant is the emotions connected to all elements of culture and, moreover, a constant dialogue among humans about these emotions as they relate to the elements of culture. These reflexive and interpersonal dynamics increase as culture is written down and passed along across generations. Distinctive ideologies and normative systems eventually develop and specify the appropriate emotions to be expressed in virtually all social situations. These ideologies and norms can even come to constitute an *emotion culture* typifying a population and various sub-populations (Hochschild 1979). Emotions are part of the human neurological system and various body systems; as such, they affect all complexes making up human nature because they are not only aroused in a biological sense but are also subject to reflection in speech acts, which, in turn, generates an emotion culture.

By viewing emotions in this way, we do not have to take a firm stand on the essentialist and constructivist arguments for the nature of emotions. I would argue that many emotions—perhaps as many as those outlined in Tables 3.2, 3.3, and 3.4—are biology based, with this base being expanded by reflexive thinking about emotions, by collective and self-discourse about emotions, and by codification of various beliefs and normative views on emotions that are part of a population's and various subpopulations' culture. Whether emotions are more or less constructed in this sense is less important than the reality that humans are the most emotional animals on earth, and it is this emotionality that makes mega societal formation possible and, alternatively, can tear such mega formations down.

Figure 8.1 on pages 179–181 outlines the evolution of the *emotions complex* as part of human nature. The inherited preadaptations and behavioral capacities and propensities inherited from LCAs of the ancestors of present-day apes and early hominins that were transformed into the emotions complex are listed on page 179 of the figure, while the outcome of their elaboration—the *evolved emotions complex*—is delineated on the following pages. In a metaphorical sense that captures the essence of the actual reality, the traits on the left of page 179 were "fed through the elaboration machine" and then came out as the complex listed on pages 180–181. All elements of the elaboration machine *are part of human nature*—that is, humans experiencing and using emotions, engaging in complex cognitive processing, using spoken and eventually written language, and codifying systems of culture. So, these components of human nature generate five *complexes* of human nature; and as I will continue to document, these complexes are interrelated and overlap, greatly affecting each other. Thus, not only did the *emotions complex* cause the expansion of the *cognitive complex*, the enhancement of human cognition had direct reverse causal effects on emotional enhancements and indirect effects through its effects on the *psychological, interaction,* and *community complexes.* Such direct, indirect, and reverse causal effects among the complexes yet to be outlined in detail can be added to those already discussed. These intersections among the complexes, however "complex," still allow us to separate to a high degree the biological basis for human nature from culture, even as elements of these complexes push humans to create the sociocultural universe. We have isolated, at least to some degree, the biology of what makes humans a unique mammal, although the great apes reveal all the traits that are the building blocks of the complexes. Moreover, it is not fully known how far other mammals in the water (e.g., dolphins, whales), on the land (elephants), and even in the air (species of parrots, macaws, and even blackbirds and magpies) have evolved. They clearly communicate with speechlike auditory signals; they organize in complex and collective ways in a fusion-fission pattern much like humans' ancestors; they are emotional and also highly intelligent with much more complex emotional and cognitive lives than we might assume; and they may even have cultural traditions that we simply do not know about since we cannot speak their language. It is, of course, likely that our nature is more advanced in the sense of producing complex social systems organized by culture more than bioprogrammers. What remains unknown is *how far along* their biology actually is. We certainly are not alone in being emotional, intelligent, and using speech and maybe even culture to mediate social relations and form social structures organizing daily activities. This knowledge can be comforting or threatening, depending on how human-centric and insecure we chooses to be.

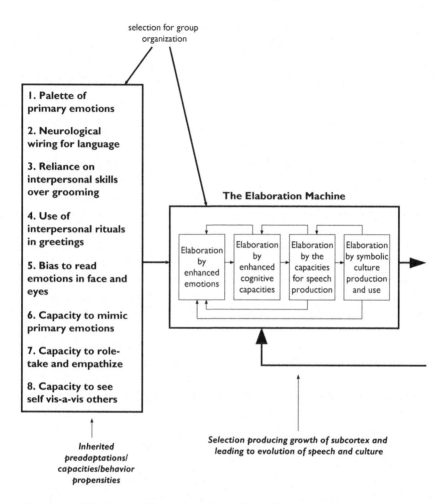

Figure 8.1 The Evolved Emotions Complex and Human Nature

1. Dramatically expanded palette of emotions with which (a) to tag ever more complex cognitions stored in short-term and longer-term memory and (b) to access alternatives in decision making, thereby making growth of the neocortex more fitness enhancing.

2. Propensity to tag with emotions all cognitions about self, others, situations, structures (and status and roles therein), social categories of persons, cultural norms, beliefs, and values, thereby moralizing virtually all dimensions of the social universe.

3. Propensity to create variants, as well as first-order and second-order elaborations (combinations), of primary emotions, with second-order variants of shame and guilt allowing for self-control as a central mechanism of social control driving human conduct.

4. Propensity to order emotions into "sentences" communicating common meanings, as a quasi "language of emotions," built up from gestures of face, eyes, body countenance, voice inflections, and other cues communicating emotional states of individuals in interaction.

5. Propensity to use speech and cultural labels in reflexive self-talk to denote new kinds of emotional states idiosyncratic to an individual but often collectively communicated within subpopulations and subcultures within a population. Such reflexive self-talk, when used to communicate emotional states to others, leads to the codification of an emotion culture among subpopulations in societies, which, in turn, leads to further talk and self-talk about emotions and moral codes.

6. Use of the "language of emotions" as the basis for fine-tuning efforts in assessing the dispositions of others, while asserting one's own dispositions, in a process of mutual role-taking and -making, status-taking and -making, identity-taking and -making, structure-taking and -making,

culture-taking and -making, situation-taking and -making, and emotion-taking and -making (see Chapter 10).

7. The propensity to build up a series of identities from mutual efforts of "taking" and "making" described in (6) above, with these identities unfolding at four levels: core- or person-level identities, categoric unit identities, corporate unit identities, and role identities. These identities consist of cognitions interlaced to varying degrees with emotions experienced during interactions with others over time, and they become powerful need-states for individuals to verify in interactions with others (see Chapter 9).

8. Capacity and propensity to repress negative emotions about self and identities from full cognitive awareness and to invoke defense mechanisms, including displacement, projection, reaction formation, sublimation, and attribution (see Chapter 3) that transmute the emotions experienced by individuals and that target others, objects, categories, and structures rather than self.

9. Capacity and propensity to recognize different levels of social structure and to experience emotions related to experiences during interactions in these structures, with these emotional experiences having large effects on identity formation and on commitments to the culture at different levels of structural organization—from groups to organizations and communities, to larger institutional systems, to stratification systems, and to societies as a whole.

9

The Evolved Psychology Complex
and Human Nature

There are many dimensions to human psychology, given the enhancing effects of subcortical and neocortical brain growth, speech, and culture, but in the end, the most relevant to understanding the biological basis of human nature are the *motivational* dimensions of human psychology. Humans are motivated to meet certain need-states, which, if not met, lead to arousal of negative emotions, a sense of deprivation, and often psychological impairment. Some of the need-states are purely biological, although they are mediated by interaction in social situations—for example, need-states for life-sustaining resources (e.g., water, food, nutrients), needs for sex, and needs for respiration and other autonomic processes. Humans and all sentient mammalian life forms share these needs as a part of their *biology complex.*

My view on need-states that drive human behavior address the same issues as most discussions on what is seen as human nature, although my portrayal is perhaps not so dramatic as some. Many of the more fundamental human behaviors that are driven by need-states are taken for granted because they are so ingrained in our nature, flying below the radar when discussing human nature. For example, I think that humans inherited what chimpanzees reveal: a hardwired incest avoidance propensity between mothers and sons, although the same hardwired trait does not exist for fathers and daughters (or fathers and sons) because it was not necessary for great apes since all females and all sons (except for chimpanzees) leave their natal locale at puberty and thus did not interact or have sex with their parents by virtue of a lack of propinquity. The same was true, perhaps even more so, for humans' hominin ancestors since males staying in their mothers' community may be a trait that evolved after the split of the ancestor of common chimpanzees from humans' hominin ancestors. If such was the case, then a biological avoidance of mothers and sons would have to evolve, as it did in chimpanzees, when sons and mothers stayed in propinquity. If, however, hominins shared the mother-son incest avoidance trait of chimpanzees, then only cultural prohibitions would reduce rates of incest in the nuclear family between fathers and offspring and perhaps sexual relations among siblings. Another sexual avoidance bioprogrammer that exists in some mammals appears to exist in humans as well: what was termed *the Westermarck effect* in an earlier chapter, named after the Finnish sociologist whose monumental *The History of Human Marriage* (3 volumes, 1891[1922] reviewed data revealing the tendency of offspring who are raised together and who engage in physical play with each other as young children to develop sexual avoidance behavioral tendencies as adolescents, thereby reducing the likelihood of brother-sister incest (see Turner and Maryanski 2005 for further discussion). Data from offspring of different families raised in the Israeli kibbutz—where parents encouraged them to engage in sexual activity as a prelude to marriage, were often disappointed because these children who were raised together in the kibbutz schools—showed the Westermarck effect (Turner and Maryanski 2005; Maryanski and Turner 2018; Maryanski et al. 2012). They did not have sexual attraction, even though there was no danger of inbreeding depression among non-siblings for offspring not sharing a high proportion of genes. Thus, other less dramatic behavioral tendencies that are part of human nature are visible but often not conceived to be part of human nature. We recognize incest taboos in human culture, and they are probably the only force working against father-daughter sexual relations (the dyad revealing the highest rates of incest in the modern family). But cultural taboos are more of a backup to the mother-son sexual avoidance (so evident between chimpanzee mothers and their sons) and to sibling avoidance once the Westermarck effect is activated in play among young siblings.

As will become clear, many treatises on human nature posit what I would see as behaviors not found among great apes. For example, sociobiologists argue for kin selection and for the view that the family is a natural social unit, driven by genetics, whereas the cladistic analysis in Chapter 2 suggests that such is *not* the case. Family orientations are better understood in terms of the emotions generated rather than by some genetic bioprogrammer that is, once again, not part of great ape biology. Other sociobiologists (Trivers 1971) and evolutionary psychologists (Barkow, Cosmides, and Tooby 1992; Buss and Schmitt 1993; Cosmides and Tooby 1992) argue that exchange reciprocity is hardwired. More support exists for this view because many higher mammals and all monkeys as well as apes evidence this behavioral propensity. So, as we will see, this assertion is more reasonable, based on some data.

Still another prominent theme in human nature arguments is that humans are driven toward dominance, toward controlling territories, toward hierarchy in general, and other behaviors revolving around power seeking. Less evidence is found to suggest that such behaviors are biology based but, instead, that they are more likely to be the result of sociocultural selection on groupings for finding a way to coordinate and control larger populations beyond the simple hunting and gathering band. These behaviors are an outcome of *sociocultural* evolution rather than biological evolution generating some need-state for hierarchy, power, and aggression; and while they are dramatic, even sexy, they are not, I believe on the basis of the data, hardwired in humans to a great degree because they are not that prominent in great apes. Chimpanzees compete for dominance when forced to live in zoos (de Waal 1982), just as humans do today in prisons, which can be zoo-like, or when chimps compete for food resources, as was the case at Gombe when the researchers started feeding animals at particular locations. They will exhibit aggression toward conspecifics in such circumstances (just as humans often do when food is thrown out of a United Nations truck to people who are starving). Thus, humans are capable of becoming power hungry and competing for resources and dominance when placed in relatively extreme conditions, but normally, they do not reveal such tendencies. Hunter-gatherers did not; they developed powerful norms against inequalities of any sort. So the potential for dominance behavior in humans likely must be activated by social, structural, ecological, or cultural conditions.

Review of the need-states that drive all human behaviors at all times show that the less dramatic ones are more fundamental to our biology than many hypothesized traits of human nature. These fundamental need-states all have something to do with two basic characteristics of hominin and then human evolution: (1) weak-tie propensities to form relations coupled with a high degree of individualism, and (2) selection pressures to get such individualistic ape-like animals better organized. Selection worked on what it could, mostly "elaborating" the preadaptations and behavioral capacities and propensities of the last common ancestors (LCAs) of great apes and hominins (see Box 2.1 and 2.2 on pages 45 and 49).

Need-States to Verify Levels of Identity Formation

The capacity for humans to see self as an object in the environment was inherited from the great apes, all of which can see themselves in a mirror. As soon as this recognition occurs, they start manipulating their gestures to present themselves in different poses, often displaying different emotions and "body language" (Gallup 1970, 1979, 1982). Selection may not have had to select directly on such phenotypical behaviors driven by specific neurological nuclei, if they actually exist as discrete structures; rather, selection that enlarged the subcortex and neocortex produced emotions and cognitive capacities for identity formation as a by-product of making hominins more emotional and more intelligent. If identity formation turned out to enhance fitness by increasing social ties and group solidarities, then selection would push for what eventually evolved: an animal capable of having a number of emotionally charged identities that would give predictability to behaviors. The end result was animals that, in virtually every interaction in every situation, are attuned to how others perceive their self-presentations. These self-presentations are organized by at least four fundamental types of identity formation: (1) *person or core identity*, (2) *categoric-unit identity*, (3) *corporate-unit identity*, and (4) *role identity* (Turner 1988, 2002, 2007, 2010b; see also Burke and Stets 2009). There can, of course, be more potential identities; and one can have multiple categoric-unit, corporate-unit, and role identities. Typically, in my view at least, people have only one person-level or core identity, but the existence of pathological syndromes of multiple personalities indicates that humans can possess multiple person-level identities. Although some identity theorists would disagree (e.g., Burke and Stets 2021), I see identities as becoming need-states for verification and confirmation by others, with individuals experiencing positive emotions when they perceive that others have verified an identity, and conversely, experiencing negative emotions and often activating defense mechanisms when identities are not verified by others (based in Cooley's emphasis on self-appraisal as arousing low-level pride or shame and in Mead's view of self-verification as an "impulse" driving the phases of the act). The configuration of emotions that individuals experience over time in this process of seeking identity verification becomes a part of an identity and, thereby, determines persons' feelings about themselves, which, in turn, affects their motive states at any given time.

Person or Core Identities

The need for verification of *core or person identity* is probably the most powerful need-state because this identity organizes fundamental feelings and cognitions about who a person is. Although emotions surrounding the core identity are often vague and transmuted by the operation of defense mechanisms and by potentially conflicting cognitions about a person's character, these often convoluted emotions and cognitions still generate needs for verification of an individual's person identity in most social situations, particularly those involving others who are important to the individual. Humans normally have trouble articulating the exact nature of their core self or identity, but their emotional reactions, whether positive or negative, will generally reveal the degree to which a person implicitly perceives that a core identity has been verified or has failed to be fully verified by others.

Moreover, following both George Herbert Mead (1938) and Sigmund Freud (1899), core identities can be built around both conscious and repressed impulses that have never been consummated. As both Mead and Freud emphasized, impulses that remain unconsummated persist and often intensify because of failures at verification when interacting with significant others. For example, as previously noted, a boy who has never received love from a father may well have repressed the negative emotions arising from this sense of rejection, but they nonetheless come out in a transmuted form as chronic need to seek approval from his father or, through reaction formation, as excessive positive feelings about his father (see Table 3.5 on page 79). Because core self or identity is codified by the *accumulated experiences* over a person's lifetime, this highly generalized identity is almost always composed of dissonant cognitions and conflicts among emotions. Coupled with repression and the activation of defense mechanisms, it is normal for a person to not have full access to the cognitions and the emotions from which this core identity is constructed. Nonetheless, it is a powerful motivational force of all *Homo sapiens*, and it is the driving force behind much thought, reflection, emotional arousal, behavior, and interaction in virtually every situation. Individuals' implicit sense about the degree to which this level of identity has been verified is determinative of not only their emotional states but also their cognitive states as they generate a sense of well-being or, alternatively, various levels of despair.

Categoric-Unit Identities

Categoric-unit identities are the next most general level of identity, with failure to verify this identity also leading to negative emotions that often get repressed and transmuted in complex ways (see Table 3.5 on page 79). Categoric-unit identities are built from cognitions and emotions related to salient distinctions among persons evident in a population. Identities form around universal social categories, such as sex, gender, and age, and often become part of a core identity, even while exerting an independent force as a categoric-unit identity or sets of categoric-unit identities, such as identities revolving around one's ethnicity and social class. Beyond age and gender, categoric-unit identities can emerge around ethnic/racial distinctions that are salient among members of a population, and still others can revolve around religious categories, social class categories, and just about any human trait that marks persons as "different" from others (recall, as part of the cognitive complex, humans categorize by contrast conceptions as one form of constructing cognitions and emotions attached to cognitions).

Categoric-unit identities are greatly affected by the degree to which such identities are valued or stigmatized by the cultural beliefs of a population. Individuals often have little choice about how they see themselves as members of a valued or devalued social category and may show elements of stigma and the negative self-feelings about the categories to which they are assigned. Children are generally treated in a positive way, but once a child becomes older and begins to receive information about those traits that are devalued—whether gender, sexual preferences, or ethnicity and related "racial" traits like skin tones—this emerging adult must now cope with potential negative evaluations by others outside his or her supportive family, although negative judgments can be avoided for a time if interaction with those who engage in stigmatizing can be avoided. Discrimination by class, ethnicity, race (appearance), gender, sexual orientation, religion, and other salient categories in a population can also be compounded or consolidated, as is the case when someone is poor, incumbent in stigmatized lower class(es), a member of a stigmatized ethnic/racial category, and a member of the more stigmatized gender. Compounding discrimination forces a person to absorb negative evaluations in many more situations, resulting in a chronic state of failing to meet needs to have categoric-unit identity verified in a positive way.[1]

In the simple societies in which humans evolved, equality among the most salient social categories (age and gender) worked to allow individuals to verify their categoric-unit identities in a positive way. As societies grew and became more differentiated, and as migrations of new categories of people across societal boundaries generated more diversity, categoric-unit identities could increasingly pose problems. Meeting the need-states of human nature to experience positive emotions can become difficult in societies where inequalities in the evaluation of categoric-unit memberships exist. People often have to accept negative evaluations of who they are, and this evaluation will also penetrate core identities, setting up conflicts in cognitions and emotions about self. People are guided more by emotions than by cognitions, and hence, they want to experience positive emotions about their identities, even if evaluated negatively by cultural beliefs. For example, a society devaluing dark-skin persons will reveal cultural beliefs emphasizing this fact, leading to a conflict between what is possible in such a society for a dark-skinned person and what their need to experience positive emotions in all situations would dictate. Persons in this situation are thus stuck with a chronic incongruence at the level of their motivations about the most important dimension of human psychology: identity.

Corporate-Unit Identities

The next level of identity formation is a *corporate-unit identity*.[2] These identities are tied to social units that reveal divisions of labor at several different levels: *groups, organizations,* and *communities* (Turner 2013). People can potentially have identities tied to those specific groups in which they participate, such as a social club or nuclear family, organizations revealing a division of labor to achieve some goal (business, military unit, school, church), or a community where one lives or once lived (town, city). These group identities can be of low or high salience to a person and can shift during a person's life course as they move into and out of various groups, organizations, and communities. Corporate-unit identities are generally less permanent and often less salient but, to the degree that they are important to a person, they create an identity that individuals seek to verify within and outside the corporate units to which this identity is attached. These identities are typically easier to have verified because many of those providing the verification are also incumbents in the same corporate unit. Corporate units themselves can be negatively evaluated, however, in which case verification of a person's identity carries stigma to those outside the corporate unit. Thus, members of a street gang are likely to verify in a positive way, among themselves, their mutual identities by virtue of their respective memberships in this type of corporate unit. Yet, others in the community where the gangs reside and operate may offer mixed evaluation, if not predominately negative evaluation, of "gang bangers." In fact, even if the rest of the society may offer only negative evaluations, these evaluations are viewed by gang members as less salient because gang members do not have to interact with members of society who impose negative evaluations. In fact, much of the draw of members in the recruiting efforts of gangs is the capacity to provide consistent verification to stigmatized categoric-unit identities, such as those related to class locations and ethnic traits that are devalued in a society. Thus, the gang-level corporate-unit affiliation offers an extra inducement for a member to adopt a corporate-unit identity held by members of the gang. Similarly, some may stigmatize corporate-unit identities tied to organizations (e.g., the military, political party, social movement organization) or communities (e.g., poor, overloaded with stigmatized members of categoric units); and under these conditions, those with stigmatized identities will tend to limit their interactions to those others who are willing to verify in a positive way their corporate-unit identity. Such selective interactions thus operate as a kind of interpersonal defense mechanism to protect self from negative evaluation, but these selective interactions also contribute to divisions and stratification in human societies.

Role Identities

A fourth level of self is *role identity*, which is an identity tied to the role that a person occupies in the division of labor of a corporate unit. This identity may also, of course, be tied to a corporate-unit identity, as would be the case of a professor (role) in a particular university (corporate unit). For some, it is the role of professor that is more salient, whereas for others it might be the university affiliation, and probably in many cases, it is both. The relative emphasis on the role over the corporate unit in which this role is played becomes clear in how a person presents self to others outside the corporate unit. For example, during most of my career, I was more concerned about my role as a publishing university professor and only secondarily about the university in which I played this role. If asked what I did for a living, I would always say "I am a college professor" rather than "I work at U.C. Riverside"—a small difference in emphasis that tells a tale about my identity. In retirement, I am nominally affiliated with two universities (U. C. Riverside and U. C. Santa Barbara), but I am not a professor in either in any real sense. I do not teach or do university service anymore, and hence, this role identity is not as relevant, nor is my university affiliation and corporate-unit identity. Thus, a new role identity is emerging through spending a majority of my time as an academic writer or professional sociologist, or some such thing. Thus, of all identities, role identities are the most numerous, pliable, and often shifting.

As is the case with my new role identity as an academic writer (who is rarely on my universities' campuses), there can be many more potential generalized role identities not attached to a corporate unit that people feel are salient and important, and thus require verifying responses from others. Moreover, these identities can become conflated with each other.

For example, a person's religion is potentially both a categoric-unit identity (e.g., a Muslim, Christian, or Jew) if religious affiliation is treated like gender, race, or ethnicity, but a devout member of a religious corporate unit like a mosque, church, or temple is also a member of a corporate unit and, hence, will have a corporate-unit identity associated with religion as well as a salient role identity within this corporate unit.

Moreover, people may assume and expect others to verify other highly generalized role-identities. For example, always being nice and friendly, rational and controlled, aggressive, shy and withdrawn, hyper and excitable, ethical and moral, responsive and caring, empathic and concerned, cool and calm, and so on are all generalized role identities that people often assume and try, if they can, to act out in all or many different types of situations. More specialized and yet generalized roles can be played out in diverse situations—role identities such as host, caregiver, partygoer, friend, companion, confidant, customer, client, and so on that people know how to play and normally can adjust to requirements of these specialized but still generalized roles in varying types of situations. Some role identities, such as father, mother, grandfather, and grandmother, that are also tied to corporate units can take on a more generalized mantra, as is the case when someone is described as "motherly" or "grandfatherly" outside their respective nuclear families. Indeed, I feel quite "grandfatherly" to my nine grandchildren but also to other people (neighbor's children, former students, friends who are decades younger), especially as I get older. I have thus conflated (a) my categoric-unit identity revolving around age with (b) style of interpersonal demeanor and (c) kinship relations into a generalized role identity.

The Loose Hierarchy of Human Identities

Core identities are always composed of elements of all other identities (i.e., categoric-unit, corporate-unit, and role identities plus any additional identities a person habitually presents to others). These identities can be seen as a hierarchy in terms of the intensity of needs for verification, with core identity generating the most powerful need-states, categoric-unit identity the next most powerful, corporate-unit identity the next most salient needs, and role identity the least salient need-states. However, individuals who are members of devalued and stigmatized categoric units may downplay verification of this identity, especially when they can more readily have corporate-unit and role identities verified by others. Alternatively, they limit their interaction to only those that verify the identity, if they can. The amount of emotion attached to an identity generally varies, with emotional intensity increasing in movement up the hierarchy from role identity through corporate identities and categoric identities to core identity. Conversely, the clarity of cognitions about identities is generally the most confused and conflated at the top of the hierarchy with core identity and then increasing in clarity with movement down the hierarchy. Where emotions are strongest, failure to verify this emotionally charged identity causes proportionate negative emotional arousal toward self and others, often distorted or even hidden by the activation of repression and defense mechanisms. These emotional stakes tend to decline with movement down the hierarchy of identities. Figure 9.1 outlines this hierarchy among identities as part of human nature.

Ordering of identities into a hierarchy is a clear tendency, but it is not always so linear, especially when lower-level identities become particularly important to a core or person identity. Under these conditions where a role identity is central to a person's sense of who they are, failure to verify this role identity is also a failure to verify a core identity, thus leading to powerful emotional reactions that are likely to become conflated by repression and the activation of defense mechanisms. Such

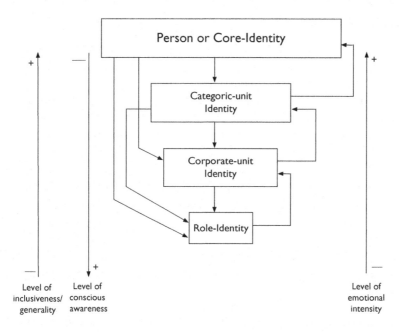

Figure 9.1 Types and Levels of Self

dynamics can be complex and often difficult to sort out. They are nonetheless an essential part of human nature. Indeed, the elaboration machine took the basic ability of hominins to see themselves as an object and converted this basic cognitive capacity into a powerful need-state capable of arousing intense emotions and potentially extreme and, at times, pathological behaviors.

Need-States for Positive Exchange Payoffs Perceived as Fair

This need-state to receive *positive exchange payoffs that are perceived as fair* is evident in the behaviors of all primates, both monkeys (Brosnan 2006; Brosnan and de Waal 2006) and great apes (de Waal 1989, 1991) as well as many other higher mammals. Humans have inherited this biology-based need-state but elaborated it in a number of ways: (a) standards of what is fair and just in exchanges of resources are often codified in cultural norms and beliefs; (b) resources that are valued are both extrinsic and intrinsic, thus making them more varied and potentially idiosyncratic among humans; (c) positive emotions aroused during resource exchanges can become an additional resource in an exchange or, if negative, a sanction or punishment imposed on exchange partners; (d) extrinsically valued resources often become conflated with intrinsic resources, especially as emotions become entangled in extrinsic resources (e.g., money is an indicator of self-worth and claims for deference and prestige); and (e) needs for verifying identities disproportionately influence exchanges among humans, particularly in defining what intrinsic resources are rewarding to persons and, moreover, in attaching themselves to extrinsic resources (when, for example, money exchanged is seen as a sign of love and respect in a couple's marriage).

These complexities do not obviate the basic biological tendency to *exchange resources in terms of some sense of fairness or justice*, but they do make it difficult to determine what is being exchanged and the emotional reactions that humans have in exchange relations. Fortunately, humans are particularly adept at role-taking and all the extensions of role-taking (e.g., status-taking, culture-taking, structure-taking, situation-taking, emotion-taking, as will be explored in the next chapter), and may become highly attuned along many dimensions of the social universe as they navigate through these complications. Still, the basic need-state remains the same: to receive resources exceeding the cost and investment to secure these resources and to perceive/feel that the exchange has been just and fair by either intrinsic standards of the persons involved or by cultural standards of fairness and justice or both. The central dynamic is rather straightforward, but it has been significantly enhanced and elaborated beyond the basic proclivities inherited by humans from the LCAs of great apes and hominins to engage in fair exchanges.

Need-States for a Sense of Efficacy

As evolved apes, humans are at their genetic core still highly individualistic, a trait often magnified considerably by the dynamics of identity formation, self-directed emotional states, and dramatically expanded cognitive capacities. From this elaboration, I suspect there is another emergent need-state: *to achieve a sense of efficacy*. What is denoted by this term, *efficacy*, is the ability of persons to have *some sense of control* over (a) the flow of events around self and (b) the realization of intended results or outcomes. The evolution of ever more layers of social structure and culture in human societies can often offer many diverse opportunities to achieve this sense of efficacy and, at the same time, these new layers of sociocultural formations can operate as a constraint on human needs to operate as individuals and to exert some control over outcomes of behaviors. For example, Karl Marx's basic argument about alienation was, in fact, an argument that it is human nature to want to have control over our acts of production (i.e., what one produces, how one goes about the tasks of producing, and to whom one distributes the products of specific labor). Max Weber's concern over the Iron Cage (better with a more accurate translation: "steel enclosure") of rational-legal bureaucracies on individuals is also implicitly indicating that humans need to feel a sense of self-control over their actions and outcomes of these actions. Herbert Spencer's concern over the constraints imposed by consolidation of power and authority, as well as the constraints of stratification, is a similar argument. The entire corpus of utilitarianism and neoclassical economics is, of course, based on the view that open, free, and unregulated markets give individuals the best chance to pursue maximization of utilities by virtue of their capacities to be rational and thereby make their own decisions about their best courses of action.

Humans can adapt to highly restrictive, constraining, and even oppressive sociocultural systems, but this need to experience a sense of efficacy operates as subtle pressure against such constraints. Humans are evolved apes, not evolved monkeys, and thus they prefer, to some high degree, to make their own way. Humans can tolerate some constraints, as long as they can achieve efficacy in other pursuits, especially those revolving around exchanging valued resources and verifying important identities. As Georg Simmel (1906) argued, market-driven, highly differentiated societies offer many options for individuals to pursue diverse lines of action, mediated by markets; he felt that such systems opened up opportunities, as did Spencer (1874-1896) and Durkheim (1893), as did even Marx (except he saw not enough equality in opportunities in capitalism as well as the tendency of ever more people to be driven into the proletariat; hence the need for his form of socialism, which, ironically and sadly, appears to have been an even more constraining force on human efficacy in its actual implementation).

It would not be possible, I believe, to understand human nature without recognizing that this need for a sense of efficacy is buried deeply in human and great ape genetics. Hunting and gathering societies were about as simple and without heavy constraints as human social organization can be. There was no inequality; resources were equally distributed within and between nuclear families, with some adjustments in the name of "fairness" for those who contributed more to acquiring these resources; no one had power over others; the division of labor (men hunt; women gather) was not a basis for gender inequalities, just the opposite since women gained prestige in the eyes of men because they brought in about 75-80% of the food; families could leave the band if unhappy; married couples could divorce; and so on. Individuals were relatively free to do as they wanted within the basic constraint of getting enough food and shelter to survive. Societal evolution began to take away such opportunities for being free and efficacious. Still, even as the populations of societies grew, leading to the organization of the economic activity and distribution of power first within unilineal descent kinship systems and, later, as bureaucracies of emerging states extended coercive control, new opportunities arose with urbanization and the evolution of market systems, although these transformations of societies also generated high levels of inequality, reducing the sense of efficacy for many. Indeed, the growing level of stratification imposed a restriction on ever more individuals and families, except a few elites; and so, patterns of social organization often worked against spreading the capacity to achieve a sense of efficacy. Migrations to urban centers of even restrictive agrarian societies offered some increase in this sense, as did industrialization and later postindustrialization, especially as markets became more dynamic. Do these systems provide enough channels for feeling efficacious, for verifying identities, for perceiving that individuals have control over their destinies, and so on? A lot of ideological puffery proclaims that postindustrial capitalism in a world market system does just that, but is it really true? The answer: only to a degree and only among certain segments of the population. Thus, anger may persist

across large segments of a population because one of the most fundamental needs of all humans is not being realized: to control one's destiny
in both large and small arenas. The real questions then becomes whether
it is even possible for such to be the case with large populations needing
to be organized and regulated to a higher degree than any hunter-gather.
If humans cannot return to their origins, then what will be the outcome
of this constant pressure to allow more efficacy?

Need-States for a Sense of Group Inclusion

As evolved apes, humans often experience some aversion to being engulfed in groups, particularly if they thwart meeting needs to feel efficacious and to verify self. Yet, millions of years of selection made
hominins and then humans more emotional and, through this emotional
enhancement, more oriented to stronger ties and group solidarities. As I
have argued (Turner 1987, 1988, 2002, 2007) for some time, humans
have powerful need-states for group inclusion, or a sense that one is part
of an ongoing flow of interaction now and in the future when the interaction is iterated. This need for a sense of group inclusion does not
necessarily mean that one must be in high solidarity with a set of others
in every interaction. Rather, all that is often necessary is for self to feel
accepted by others and part of the ongoing flow of the interaction. As
humans inherited and then magnified capacities to engage in emotion-
arousing rituals and in synchronized rhythmic flows of talk and bodies,
this need for inclusion in the ongoing interaction can also lead to enhanced solidarities over time, as the interaction is strung out in chains of
interaction among the same individuals (Collins 1975, 2004). If positive
emotions are consistently aroused, then other need-states for humans are
also met.

Evolved needs to verify levels of self, to achieve profitable and just
exchange payoffs, and to feel a sense of efficacy mean that humans are
always seeking to experience positive emotions related to self that may
come into conflict with those group inclusion needs to feel part of the
ongoing interaction—the very issue that Durkheim was trying to understand in his types of suicide—*anomic, egoistic, altruistic,* and *fatalistic.*
A delicate balance is required for an individualistic ape to avoid feeling
engulfed by groups and other social structures that may thwart individualism, verification of identities, and needs for efficacy. Once more
permanent groups emerge as a "survivor machine" (Dawkins 1976), the
need for group inclusion that emerged along the human line and interactions were increasingly constrained by group affiliations. Individuals

will generally tolerate some constraint by groups as necessary, if they are still allowed to meet most other psychological need-states in the psychology, emotions, and interaction complexes. Therefore, group memberships and activities that do not allow a sufficient portion of need-states to be consummated will reveal tensions and negative emotional arousal.

In the present world, these escalating needs for a sense of group inclusion drives what at times can seem like compulsive behaviors in social media to feel part of a potentially large circle of "friends" or, conversely, to feel acute negative emotional combinations of fear, anger, humiliation, and hurt when "unfriended." Indeed, a dramatic rise in teen suicide rates (Abrutyn and Mueller 2016, 2018; Mueller and Abrutyn 2016) indicates how powerful the need-state for group inclusion has become, due to adolescents and others investing too many of their identities and emotional commitments in faux social universes such as Facebook and Instagram. Rejections in social media signal that a powerful but subtle need has evolved in humans to feel part of ongoing social activity and, in some cases, high levels of perceived (but perhaps illusionary) solidarity with particular sets of others. Too much solidarity can also feel engulfing and constraining because evolved apes like humans still possess ancient proclivities, inherited from their distant great ape ancestors along the hominin line, for individualism, autonomy, and mobility. The rewiring of the human brain to seek stronger social ties through emotions, and to reinforce this propensity through ritual and rhythmic synchronization (Collins 2004), was installed by natural selection alongside old propensities for some autonomy, individualism, and sense of efficacy. This potential conflict in need-states that are part of human nature is perhaps inevitable when selection forced group organization on hominins who, like humans today, are rather individualistic great apes.

Need-States for Cognitive and Emotional Congruence

A large literature in psychology documents that the human brain is wired to seek congruence, balance, and consistency among cognitions. Indeed, it may be, as I suggested in the formulation of the cognitive complex, cognitive consistency and congruence are mechanisms for ordering information in a brain now capable of holding vast stores of knowledge and information. When individuals experience cognitive inconsistency and related misalignment of cognitions, they appear motivated to bring these into line, if they can. With the need to live in a complex world and with an evolved psychology revealing potentially conflicting need-states, such is not always possible. It is, then, in the nature of humans to live with a certain amount of cognitive dissonance and incongruence. At the same time, humans are motivated to bring into line discordant information because discord among cognitions activates negative emotions, which enhances the sense of incongruence and motivates individuals to work to bring discordant cognitions and feelings into greater harmony. The evolution of repression and a series of universal defense mechanisms (see Table 3.5 on page 79) attests to this need to achieve, if only in conscious (mis)perceptions, a sense of cognitive balance. Yet, repressed emotional imbalances do not stay buried. They resurface as new imbalances activating diffuse anger, sadness, and anxiety—a situation that guarantees that humans will have to be tormented, to a degree. The social universe is now complex not only in its patterns of social organization but also cognitively and emotionally because of the need to assemble diverging and incongruent cognitions about, and emotional responses to, so many dimensions of the social universe.

As the neocortex evolved as an outcome of the dramatically increased capacity of hominins to experience ever more variants of emotions at varying levels of intensity, the brain became wired to store vast stocks of knowledge ordered in some way for retrieval by the prefrontal cortex. The key to all memory storage among life forms on earth is to tag cognitions of experiences with emotions, originally fear and anger as immediate flight-or-fight responses by mostly olfactory-dominant animals as survival mechanisms when facing danger. Mammals added ranges and variants of satisfaction-happiness and disappointment-sadness to this small palette of negative emotions. On these and perhaps a few more primary emotions, human levels of emotionality evolved. Emotions and cognitions are thus forever joined because cognitions cannot be remembered, stored, or used in fitness-enhancing decisions without being tagged with emotions (Damasio 1994). These emotional tags, if not the cognitions to which they are attached, can come into conflict, thereby creating conflicts in need-states and psychological well-being.

The brain seeks to bring the cognitions and the emotions attached to them into some kind of congruence, driving humans to reduce incongruence. There are probably many mechanisms, not fully understood, that work to this end. Defense mechanisms are one such mechanism; cognitive justifications and gymnastics can be another to bring conflicting emotions (attached to cognitions) into line, although these typically have to be accompanied by repression, transmutations, or other mechanisms for controlling the emotions involved. The dynamics of these and other mechanisms are not fully known beyond the simple insight that humans possess a persistent need to align cognitions and emotions attached to them or to otherwise segregate neurologically incompatible cognitions and emotions.

Need-States for Sense of Trust

A suite of related need-states that humans are motivated to achieve in social relations can be labeled needs for "trust." Sociologists and philosophers have given them various labels, but I will use simple terms to describe several related needs under the heading of *trust*. One is the need to feel that others are being *sincere and honest* during the course of an interaction. Whether humans have a "cheater detection module" is less important than the recognition that humans need to feel this sense of sincerity or honesty. Humans look for signs of this feature in social relationships, which can make them vulnerable to manipulation by others who are good at mimicking signals of sincerity and honesty.

Another form of trust revolves around the need for persons to perceive that others are *respecting the salient identities* projected into the interaction. People look for signs of verification, and if they perceive that others are indeed signaling verification and acceptance of those identities, they will experience a sense of trust. When self is not verified, however, the need for trust is difficult if not impossible to achieve. For social relations to develop out of episodes of interaction, individuals need to feel this sense of trust consistently and, moreover, to feel that such will be the case into the future, and this sense depends on feeling that others are sincere and honest as well as being respectful of identities.

The last of the need-states for trust is to feel that the situation and context in which interaction is occurring *is as it seems*. Individuals often look for incongruities between what the situation is seemingly supposed to be for signs that the situation is not as it seems on the surface. This quiet "mistrust" can, of course, be taken too far, making it difficult for persons to ever feel that others, situations, and interactions are as they seem. Paranoid persons are thus almost always ill-disposed to meet needs for trust, primarily because their underlying anxieties and other forms of repressed fear, often mixed with anger, make it difficult to accept situations for what they are. Even situations that are not critical or essential to their well-being can be subject to a lack of trust, thus making the life of a paranoid miserable and, in extreme cases, highly dysfunctional.

Thus, this set of need-states for trust is a complex mix of potential problems in an interaction and in social relations. Individuals monitor others for signs that they are not being sincere and honest as well as respectful of self and identities now and, by inference, into the future. A sense of trust is always coupled with an implicit search for signs that others are disingenuous, self is not being respected, and the situation and events transpiring are not "as they seem" on the surface. These three elements in the need for trust are what con artists emphasize—"I am sincere, I respect you, and everything is just as it seems"—as a simple interpersonal three-part strategy to gain trust.

Humans vary from being gullible in a rush to meet these needs for trust to being habitually cynical and even paranoid with anxieties and fears leading to chronic mistrust. Most humans fall safely between these extremes, but nonetheless, this need for trust is directing their thoughts and actions, at least until a sense of trust is firmly established, which means that, for the con artist, this is the moment to strike.

Need-States for Experiencing Positive Emotions

All these need-states are built on a meta-need for humans to *experience positive emotions* over negative emotions in all situations, especially in regard to a person's various levels of identity. People will, of course, endure situations where they are experiencing negative emotions if they perceive that these are simply normal costs to greater emotional payoffs involving positive emotions. Sometimes such perceptions can be illusionary or even a defensive mechanism for enduring negative emotions. For the need-states examined in this section, meeting needs for verification of identities, needs for positive and fair exchange payoffs, needs for a sense of efficacy, needs for group inclusion, needs for cognitive and emotional congruence, and needs for trust are all part of a larger need to experience positive emotions, which, of course, is impossible for every human all the time, and particularly in big, complex, and constraining societies. Furthermore, even after millions of years of evolution, the palette of human emotions is still weighed toward the negative. Indeed, humans are well aware of the fact that they are susceptible to many negative emotions because there are simply more negative than positive primary emotions in the human palette, and these inevitably get activated in at least some situations.

To interact with others, humans must be able to read, experience, and express negative as well as positive emotions; and they have a large repertoire to draw upon. Such is the cost of early wiring of the amphibians that spawned mammals and reptiles with a hair-trigger amygdala generating fear and defensive anger as the immediate reactions to danger. Without these emotions, there would be no mammals or reptiles living today.

Positive emotions among mammals began to evolve for nurturing dependent infants and later from selection on subcortical areas of the brain for producing hormones, neurotransmitters, and neuroactive peptides to accompany bioprogrammers for kin relations to counter raw fear and anger. Play among young mammals is perhaps one early path through which positive emotional development began (Huizinga 1938 [1955]; Burghardt 2005; Deacon 2009; Bellah 2011), supplementing the positive emotional attachments of mothers to their offspring. For hominins, natural selection worked for millions of years to dramatically increase the variety, types, and intensity of positive emotions beyond that of other mammals, but these evolutionary processes started with a larger palette of negative emotions revolving around fear, anger, and sadness that had to be mitigated and supplemented by positive emotions. It is for this reason that the *amygdala* is larger in humans, with nuclei, especially the lateral nuclei, that can produce more positive emotions in the ancient area for fear and anger (Barger et al. 2014). Selection worked to increase positive emotions for strengthening of social ties and group solidarities because it was the only path that natural selection could take for increasing sociality and groupness among weak-tie, low-sociality great apes that needed to be better organized if they were to survive on the savanna and other open-country habitats.

The end result is what we see today in the human world: many positive emotions creating and sustaining highly rewarding social relationships and incumbency in social structures. For most of human history, humans probably experienced more positive emotions from relations in nuclear families in small bands. Yet, neurologically, humans have always been capable of intense negative emotional arousal that can, at times, work to move us out of environmental threats and danger but that can also disrupt social relations, group solidarities, commitments to others, and even entire organizational patterns of sociocultural formations. As societies eventually began to grow, as inequalities increased, as warfare among populations became chronic, and as other features of societies caged and exploited humans in social structures (Maryanski and Turner 1992), this potential for extreme negative emotional arousal was activated. Even with these obvious stimuli for negative emotional arousal, however, societies and their constituent structures can only be sustained and reproduced by the bias, and indeed the need, that humans have for experiencing positive emotions. Humans are constantly seeking positive emotional arousal in as many situations as possible. This need to experience positive emotions works to keep the dark side of humanity sufficiently at bay for most people, although all humans today must live with negative emotions, and some must endure negative emotions more than positive emotions in their daily lives. Still, the positive emotions are what keep humans going, reproducing the species in kin units where positive emotions are more likely to prevail and in social groupings in communities and organizations that can bring other sources of positive emotional arousal. The balance between positive emotions promoting sociality and solidarity faces a host of situations in the modern world that generate negative emotions, eroding solidarities and humans' sense of well-being. Fortunately, humans' powerful needs to experience positive emotions and, to some degree, avoid negative emotions, are perhaps the salvation of the species in the modern world.

Conclusion: The Evolved Psychology Complex

Figure 9.2 on page 203 delineates the process of elaboration on the inherited traits from the LCAs of the ancestors of great apes and hominins. The preadaptations and behavioral capacities and propensities on the left of the figure are what were fed through the "elaboration machine." These traits were also affected by what the elaboration machine had already done to the other complexes, particularly the emotion, cognitive, and interpersonal complexes. Once changes in one complex have been installed by natural selection, they have direct and reverse causal effects on the other complexes. Thus, the motive states of human psychology are not just psychological; they have also been affected by changes in the other complexes, which makes human nature that much more complicated. Humans are not driven by a few signature motivating traits, such as needs for dominance, for protecting kin, for power, for any of the many single behavioral tendencies we might mention. Rather, human nature is a complex of complexes, if this makes any sense, which are somewhat less dramatic but nonetheless fundamental because they make us humans who we are and drive, constrain, and direct human actions.

All these traits are the result of selection pressures for stronger social relations and group solidarities, but as selection worked in this direction, it created complexes that significantly went beyond group-level organization. Selection installed a complex set of traits that would take a marginally successful primate and convert this primate into an animal that would eventually dominate the planet, for better and for worse. Thus, once these complexes exist and begin to interact with each other under varying selection pressures, they produce *emergent properties far beyond what might be predicted.* Such is the nature of selection as it sets into motion synergies among traits—some dormant as preadaptations, others seeming unimportant until selection enhances them, and all subject to transformation as the elaboration machine does its work in expanding and intensifying them even further. The outcome is the evolution of a species that might not have easily been predicted, if we had the luxury of watching the evolutionary process unfold over 5 million years. We have been able to get a peek at these processes through cladistics analysis, comparative neuroanatomy, and the direction of selection driving hominin evolution, but obviously this picture remains picture with only somewhat limited abilities to bring into focus each evolved complex under consideration—in this case, *the evolved psychology complex.*

Big-brained animals that have been supercharged with emotions and evolved cognitively to the point where speech and language can produce culture will have virtually unlimited needs and desires, some shared with other humans but most unique to particular individuals and their life experiences. That said, my concern is what *all humans share*, as a species, rather than the diversity and uniqueness that an animal like humans will reveal. So the list of motive states that typify all humans may not seem as dramatic as many other treatises on human nature, but may perhaps be truer to what humans are. What we are, however, has allowed humans to create forms of sociocultural organization that create new pressures on this evolved neurology. The result has been the evolution of a series of sociocultural formations for human organization over the past 12,000 years that have often violated elements of humans' evolved nature. To appreciate this fact, it has been necessary to separate the biological from the sociocultural, as best we can, and then look at the biological and ask, To what degree do the various forms of social organization created by humans violate or accommodate humans' evolved complexes? Clearly, the answer is that the organization of societies more generally has not accommodated, and indeed violated, much of human nature, although humans as adaptable animals have found niches and places within patterns of social organization where their needs and other traits of human nature can be accommodated. Before making any judgment, we need to examine the two more complexes: the *evolved interaction complex and the evolved community complex.*

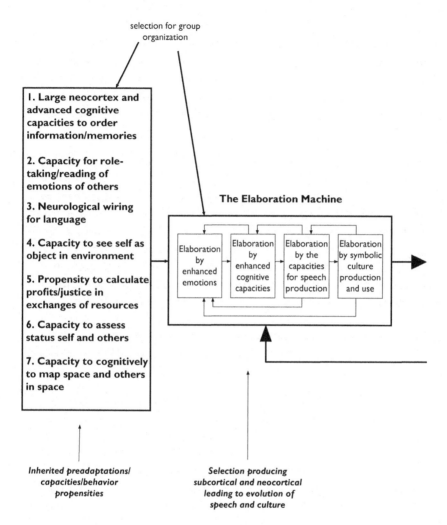

Figure 9.2 The Evolved Psychology Complex and Human Nature

1. Propensity to develop identities forming at least four levels: (a) core or *person-level identities*, (b) *categoric-unit identities*, (c) *corporate-unit identities*, and (d) *role identities* generating. These identities are motivational need-states, with person-level being the most powerful of these need-states.

2. Clarity of cognitions, intensity of emotions, and operation of dense mechanisms vary across types of identity. Core or person identities are the most emotionally loaded and conflated with the operation of defense mechanisms and, hence, the least subject to conscious awareness; categoric-unit identities are the next most conflated; corporate-unit identities and role identities are less emotionally infused, less subject to operation of defense mechanisms, and most subject to cognitive awareness.

3. Person-level and categoric-unit identities are the most stable over time, while corporate-unit and role identities can change over the life course.

4. Identities that are not verified, or negatively evaluated by cultural beliefs, arouse anger and other negative emotions toward others or, alternatively, arouse *shame* or *guilt* leading to the activation of defense mechanisms. Humans can often selectively present only those identities that can be verified in a positive way in order to avoid the negative emotions and potential activation of defense mechanisms.

5. Humans have needs to experience (a) receipt of resources more valued than the costs in interaction with others, and (b) receipt of resources that meet or exceed cultural and personal standards of justice. Failure to experience (a) or (b) arouses intense negative emotions and will lessen commitments to others, situations, and structures in which this failure has occurred. Conversely, profitable and just exchange payoffs lead individuals to experience positive emotions and increase their commitments to others, situations, and structures.

6. Humans have needs to experience a sense of efficacy in behaviors and the outcomes of behaviors, with failure to realize this need arousing negative emotions and intense negative emotions if efficacy is tied to identity verification and with meeting this need likely to arouse positive emotions and commitments to others and the structure in which efficacy is experienced, especially when efficacy is connected to verification of identities.

7. Humans have needs for a sense of group inclusion or being part of the ongoing interaction, and especially so if an identity is on the line, with individuals experiencing positive emotions and commitments to others and social structures when this need is met and negative emotions and lowered commitments when this need is not met.

8. Humans seek, when they can, to experience balance, congruence, and consistency among cognitions and emotions, which, if associated with cognitions and emotions about identities, arouse the same dynamics as outlined in (4); if not conflated with identities, the emotional reaction, whether positive or negative, will be significantly muted in light of the nearly impossible task of having balance and congruence among all cognitions and emotional states.

9. Humans seek a sense of trust in all interactions, revolving around a sense that (a) others are being sincere and honest, (b) others are respecting one's key identities, and (c) the current situation is as it seems, experiencing positive emotions and commitments to others and the situation when this sense can be sustained and negative emotions when any of (a), (b), or (c) are not realized.

10. Humans always seek to experience positive emotions and avoid negative emotional arousal in all situations, with arousal of positive emotions leading to (a) commitments to others and the (b) the social unit and its culture in which positive emotions are experienced and with arousal of negative emotions leading to the reverse of (a) and (b) and, potentially, activating the operation of defense mechanisms.

Notes

1 There are somewhat contradictory data about what causes negative emotions among individuals in identity dynamics. Is any inconsistency between self and others' evaluation of an identity likely to generate a negative emotional reaction of a person, or is it only reaction to a culturally devalued identity? For example, some argue that a negative identity when given positive rewards by others will still generate negative emotions, just because of the incongruity between a negative identity and positive evaluations by others, but I would argue that people are motivated to have positive evaluations of their identities, even if they are forced to live with a stigmatized identity. Identities are not like cognitions, which generate dissonance as a motivational force. Identities are more than cognitions; they are also driven by needs for positive emotions, and, hence, even people dealing with cultural stigma for an identity want to have it viewed in a positive light.

2 In psychology, the notion of "group" identity is often used to denote what I am defining as very different types of identities: categoric-unit identity and corporate-unit identity. The former revolves around categoric distinctions that humans make, whereas the latter refers to identities build around incumbency in corporate units, which can be of three basic types: groups, organizations, or communities. Lumping these various potential identities under the label *group* obscures significant differences that need to be more explicit to understand identity dynamics. Furthermore, biologists, psychologists, and even economists often use the term *group* to denote any pattern of social organization, which conflates many different forms of sociocultural organization.

10
The Evolved Interaction Complex and Human Nature

The emotions, cognitive, and psychological complexes outlined thus far constitute three of the most critical dimensions of human nature. Yet, these complexes must be activated in infancy by interaction with other humans. The emotional, cognitive, and psychological potential of humans thus requires that infants live in a speech environment where emotions are communicated; and the interaction and community complexes are the same: they require activation when a child is an infant and, if not activated in this normal manner, a child at 10 or 11 will not only have trouble learning how to speak, but also the child's emotional range, cognitive abilities, capacities for interpersonal behaviors, arousal of psychological motive states, and sense of the sociocultural universe will never be developed fully. Such a child will appear odd and, if deprived for too long, will appear not to be human in its basic nature. For example, children living with wolves become more like wolves when their first interactions are with wolves rather than humans. Thus, human nature is not inevitably stamped on humans; human nature is bundles of capacities and propensities for behaviors that must be activated by interaction with others living in sociocultural formations. The fate of feral children is illustrative, as is outlined in Box 10.1.

Elements of the Evolved Interaction Complex

The most unique features of great apes and, no doubt, humans' common ancestors with extant great apes are their individualism, weak-tie relations, lack of permanent groups, sexual promiscuity, and weak kinship system. All these, as indicated in Chapter 2, are the outcome to having lost out in the competition to group-oriented monkeys in the arboreal habitat in Africa, forcing the ancestors of great apes to adapt to the marginal niches in the terminal feeding areas of trees where there is not enough room, structural support from branches, or food to support permanent and large groupings. Hence, the system of male and female transfer away from their mothers (and from their unknown fathers because of great ape promiscuity) evolved to disperse adolescents at

BOX 10.1 *FERAL CHILDREN:* DO THEY COME TO
POSSESS HUMAN NATURE?

In 1920, the Rev. L. Singh confirmed a rumor in rural India that there were ghosts residing in an anthill. After setting up an observation post, he convinced workers to open up the ant hill, and they were immediately attacked by a mother wolf, which they immediately killed. To their surprise they discovered that the mother had left behind four of her charges: two wolf cubs and two little girls, one about eight years old and the other about 1.5 years old. The girls were "wolfish" in appearance and behaviors; they had hard calluses on their knees and palms from walking on all fours; they moved their nostrils to sniff food. Eating and drinking involved lowering their faces into the food. They ate raw meat and hunted wild animals. When brought back to human society, Kamala and Amala shunned other children and preferred the company of the dog and cat. When sleeping, they rolled up together on the floor. No one knew how the girls came to be there, but one thing was clear: they acted more like wolves than humans.

Thus, human nature has to be activated by humans and by interaction with humans. If your only interactions are with wolves, then you begin to develop wolf nature. Other cases of feral children are equally illustrative. The case of "Anna of the Attic" provides further information about what happened when Anna, an illegitimate child, was locked up in an attic with virtually no human contact, just enough to feed her. When found, Anna could not walk or talk, and in fact, because she did not respond to gestures by humans, it was thought that she was blind and deaf. With training, she overcame these debilities and was slowly learning how to communicate, but her death four years after her discovery cut short her training. Yet, it was obvious that she would never be normal.

In another case, yet another illegitimate child, Isabelle, was isolated by her mother, who was a deaf-mute; but unlike Anna, she learned to communicate with her mother through a series of guttural, croaking sounds. As a result of face-to-face interaction with her mother, and learning how to communicate in a simple way, much of her nature as a human was activated. The result was that later she was able to become almost, but not quite, normal after training.

Clearly, human biology does not ensure that one becomes fully human. If you are raised by a wolf, you begin to act like wolf, not a human; if you are isolated, you become unable to walk or talk and do not learn how to fully be human. If, however, you see your human mother face to face, even though she cannot hear you or speak to you, just enough is activated by this interaction to enable the child to create her own language for communicating and, later, to learn the natural language of her society. With interaction, then, the evolved emotions, cognitive, and psychological complexes can be activated.

puberty to new areas of the arboreal habitat. Unlike most other mammals, then, the ancestors of contemporary great apes and humans evidenced mostly weak-tie relations, no nuclear families, no bioprogrammers for kin selection (beyond mother–infant/young offspring bonds), no permanent groups in general, and an orientation to a larger community of conspecifics rather than to local groupings.

Because early great apes were probably much like the orangutans of today (see Table 2.1 on page 35), they evolved a set of interpersonal practices that allowed for them to function in weak-tie relations. These were inherited by humans, even as natural selection randomly searched for ways to make hominins more social and group oriented, as was emphasized in Chapter 5, where the inherited interpersonal facilities enabling late hominins and early humans to forge flexible group structures were examined. The elaboration of the inherited interpersonal abilities of the last common ancestors (LCAs) to present-day great apes and early hominins was made possible by the evolution of the emotions, cognitive, and psychological complexes, but as natural selection began to enhance the interpersonal capacities of hominins, these enhanced interpersonal capacities had reverse causal effects on other complexes comprising human nature, which, in turn, would lead to further expansion of hominin and then human interpersonal abilities.

Few mammals can do something that is simple for humans: interact with strangers without stress (Moffett 2018). Most mammals cannot walk through crowds of unrelated conspecifics, stand in lines of non-kin whom they have never met, travel enclosed on trains and trolleys packed with conspecifics whom they do not know, walk into buildings where they have never been and engage in necessary interactions, and do all the many interpersonal gymnastics that humans see as normal and natural. Humans can engage these interpersonal practices because, once activated in infancy and childhood, they become part of humans' evolved nature. Compared to other mammals that are often skittish in unfamiliar settings with conspecifics that are not members of their local groupings, humans' interpersonal abilities are rather remarkable. These evolved interpersonal skills can be seen as an evolved *interaction complex*. The most important elements of this complex are examined next.

Evolved Capacities for Identity Formation and the Presentation of Self

Probably the greatest elaboration of humans' interpersonal abilities and propensities is related to the presentation of self in social relations. As emphasized for the cognitive and psychological complexes, humans have the capacity and, indeed, the propensity to form multiple levels and types of identities that are highly associated with the emotions experienced in their development during the course of the life cycle. As a result of this new capacity to acquire multiple levels of self, humans invoke identities in virtually all episodes of interaction in all social contexts, not just as a capacity but as a powerful need-state. Thus, active *role-taking* and *role-making*, which are the essence of all human interactions (Mead 1934), are guided by identities in the sense that individuals (1) mutually present an identity or identities to each other, (2) assess the implicit or explicit evaluation of this presentation by reading the gestures of others for the degree to which they accept this identity presentation, (3) experience positive emotions when an identity is perceived to have been accepted by others and negative emotions when identities are perceived as not being accepted by others, and (4) activate defense mechanisms when an identity is not accepted and/or express negative emotions toward others. Thus, among humans, the interpersonal flow revolves much more around *mutual verification/nonverification of identities* when compared to great apes and, hence, early hominins. Yet, great apes have some capacities along these lines. For example, pictures on the Internet show "standing gorillas" who are members, and consider themselves as members, of an enforcement organization that is trying to stop poaching on gorillas. Their human "colleagues" are standing around in their compound with these large gorilla "colleagues," who also stand erect. They stand this way because they see themselves as humans, and humans stand and walk bipedally. Clearly, their bipedalism represents some sense of an identity of themselves as human or almost human. Similarly, the picture-sorting task assigned to great apes raised with humans, who place the picture of themselves in the pile of human photographs rather than the pile of ape pictures, suggests that they "see themselves" as human because they have been raised by humans. Moreover, as already mentioned in earlier chapters, perhaps even more dramatic is the propensity of an adolescent female chimpanzee raised with humans to be more sexually attracted to, and flirtatious with, young human males over available chimpanzee males. Again, a more human identity has been formed even around something as biologically fundamental as sex. Thus, the human propensity to form identities is clearly inherited from great apes, who, when in a human environment, begin to form humanlike identities, perhaps like the wolf children in India who act more like wolves than humans.

As was outlined in the previous chapters on the cognitive and psychology complex, humans have the cognitive capacity to develop emotion-laden conceptions of themselves as certain types of persons and beings deserving of particular responses from others on at least four levels. The top level of self is *core or person identity* and is built up over a lifetime by emotionally charged cognitions about who a person is. Verifying a person-level identity is one of the most powerful motivations that humans reveal. Because verification of this identity is so important, individuals engage, as I will later examine, in a process of *identity-taking* (reading the gestures of others for the identity being presented) and *identity-making* (orchestrating gestures to communicate an identity for others to verify).

A second level of self are *categoric-unit identities* that are less general than core or person identities, but they are still general because a person typically must carry some of these categoric-unit identities, such as sex and gender, for a lifetime—although these too can be altered. Moreover, categoric units are subject to differential evaluation by cultural beliefs—what are often termed *status beliefs* by sociologists (Berger 1958, 1972, 1988; Berger et al. 1977; Berger and Zelditch 1985; Ridgeway 2001). These evaluations of members of a population of different categoric units are inevitably applied to identities built around categoric-unit memberships. Such evaluations are generally part of others' evaluations of categoric-unit identities being presented by a person, as well as being part of this person's – self-feelings about the worth of their categoric identity in the broader society.

Sometimes the evaluations of persons' multiple categoric-unit memberships and the identities formed around each of these memberships consolidate in that they are all consistently high or consistently low. For example, a dark-skinned member of the lower class in a society where dark skin and poverty are stigmatized will be doubly stigmatized by this consolidation. In contrast, a wealthy, upper-class dark-skinned person in the same society evidences intersection of a valorized categoric-unit membership and its positive evaluation in cultural and status beliefs with a stigmatized categoric-unit membership and a corresponding negative evaluation in cultural and status beliefs. Intersections of categoric-unit memberships and their associated identities reduce the stigma associated with the less-valued identity because of the valorizing effects of status beliefs about the highly evaluated identity. In the example, being rich trumps, to some degree, the stigma of having dark skin in a society stigmatizing dark skin but valorizing wealth and upper-class membership. The more consolidated the identities associated with cultural evaluations of various categoric-unit memberships are, the greater will be the power of status beliefs—whether valorizing or stigmatizing—on the identities associated with these memberships (Turner 2002). For example, a white, rich and upper class, male heterosexual will receive a higher evaluation than any one of these memberships would warrant alone. Consolidation increases the evaluation of an identity in interaction beyond a simple additive effect. As a result, this person will have advantages in almost all interactions with others. Conversely, a dark-skinned, poor, and lower-class male will be more stigmatized by the consolidation of skin color, poverty, and class position and thus be at a disadvantage in many interactions with others. Being male, which in the same society may have a more favorable evaluation in status beliefs, probably will not raise the overall evaluation because of the consolidation of several highly salient stigmatized categoric-unity identities. With intersection, the negative or positive evaluations of several categoric-unit identities work at cross purposes. Hence, stigmatized identities lower the overall evaluation of a person presenting other identities that are more valorized; conversely, valorized identities will always raise the overall evaluation when presented alongside more stigmatized identities. Just how these consolidating and intersecting processes work out in actual interactions will vary depending on which identities are valorized and which are stigmatized and the salience of the valorized or stigmatized identities in a situation. The only constant is that categoric-unit memberships and identities associated with them will always be differentially evaluated. If multiple categoric-unit identities are salient in an interaction, the dynamics of consolidation or intersection on these differentially evaluated identities will have large effects on how a person is treated in an interaction and, moreover, on the likelihood that this person can receive a positive evaluation from others in identity-making or self-presentations.

Categoric-unit memberships can change, and individuals are often motivated to change memberships in order to avoid stigma associated with a particular identity, especially since stigmatized and valorized identities have large effects on how interactions play out, and increasingly so the more consolidated these identities are. People work hard to rise up the social class hierarchy; they can change religious affiliation if their current one is stigmatized, and at times, they can even change their bodies (gender, skin tone, sexual orientation) to avoid stigma (but also to become who they feel that they should be biologically). Often such changes bring more stigma than prestige, but people's willingness to change even their biological appearance is indicative of the power of categoric-unit identities in how persons are treated and, hence, their evaluation of themselves in terms of cultural beliefs about status. Because categoric-unit membership is so salient in most societies, identities associated with such memberships become a major part of a person's core identity. Interactions with others almost always reflect these self-evaluations even when, as is often the case, they are unfair and based on prejudices among members of a population. Responding to difference, however, is part of humans' cognitive nature and spills over into how they respond to "others" during interactions.

A third level of self is *corporate-unit identities*. These identities arise from memberships in corporate units, including various types of groups, organizations, and communities. Because affiliations with corporate units can change over the life course, the profile of individuals' corporate-unit identities can change. Moreover, such identities always vary in how salient and significant they are for a person's core identity and fundamental feelings about self., which can also change over time. These changes may act to alter cognitive and emotional balances in self-appraisals and, thereby, what aspects of self are operating as powerful need-states seeking verification from others. For example, identities arising around incumbency in schools, churches, workplaces, sports teams, social clubs, neighborhoods, cities and towns, and other corporate units often change in contemporary societies, whereas in early hunting and gathering societies as well as horticultural and agrarian societies, persons were incumbent in a limited number of corporate units for a lifetime, thus giving corporate-unit identities some stability and, thus, generally increasing their salience for core or person-level identities.

In contrast to categoric-unit identities that are highly visible, as is the case with age, gender, sexual orientation, and "race" (minor phenotypical differences that can have great significance nonetheless), individuals typically have more latitude in the corporate-unit identities that they present to others. In fact, when an individual tries to understand the identity or identities that another is seeking to have verified (what can be called *identity-taking*), this individual will probably be highly attuned to which of several possible corporate-unit identities the other is highlighting for verification. In general, when a person gives emphasis to a particular corporate-unit identity across different interactions (in the process of identity-making), it is likely that this particular corporate-unit identity is also a significant part of this person's core or person-level identity. Most of the time, others in their *identity-taking* efforts search for selectivity in presentations of corporate-unit identities in order to verify, if they can, this favored identity and, thereby, enjoy a successful episode of interaction. For example, in the yacht club to which I belong, the serious sailors always present themselves as members of the SBYC yacht club and never talk about any other corporate-unit memberships, even when I see them away from the yacht club. The less serious sailors, who join for the restaurant and ocean view and who do not sail or even go out in any boat at all, generally present themselves in terms of their job and corporate unit in which they work rather than as active and nautical members of the yacht club. The sets of conversations are different because, in my identity-taking, they present to me the identity that they want to be salient for our interaction—in general, I stay away from my sailing and talk about being an ex-college professor. This symmetry in our respective identity-making activities makes for often pleasant conversations as we watch the boats go by at the start of a race. When I have tried the opposite, to emphasize my sailing activities, the conversation normally does not last long and is strained.

In smaller societies, where there are many fewer corporate units with which to affiliate as well as less stratification, people are often stuck with a few corporate units—family, clubs, churches, schools (for a time), workplaces. Those that bring positive emotions in interactions with others are likely to be the ones that form a corporate-unit identity, but still, many corporate-unit affiliations may be unpleasant and, hence, difficult to build identities around. As a result, individuals will be hesitant to present identities associated with stigmatized corporate units or units in which interactions are not gratifying to others outside of such corporate units. Individuals often seek to distance themselves from corporate units that bring stigma or negative emotions via a number of strategies: hiding incumbency in these units during interactions, affecting distance and alienation from the unit, emphasizing incumbency in more valorizing and emotionally pleasant corporate units as more salient to an interaction, and even avoiding presenting any kind of corporate-unit affiliation to others.

Role identities are associated with roles that are played as part of the division of labor in a corporate unit, around which individuals may form identities (e.g., father, husband, worker, religious adherent). Individuals can play other more generalized roles and build an identity around the activities associated with a role (e.g., cook, hostess, party animal, activist, iconoclast, friend, socialite, philanthropist, upbeat, alienated). Often people more strongly identify with roles in corporate units over the unit itself, or roles that are played outside of corporate units or, at least, are not confined to any one type of corporate unit. As a general rule, as corporate-unit affiliations decline or are limited, and particularly if they are ungratifying, humans seek validation for who they are through more generalized roles revolving around particular syndromes of behaviors and demeanors. Thus, it is in *the nature of humans to seek out roles that individuals can play, build an identity around, and display to others and receive some validation of their competence in this role.* If corporate-unit roles are not easily available or gratifying, and if corporate units are not sufficiently valued in a society and/or pleasant to be incumbent in, then people are more likely to gravitate to more generalized roles. The same is true if categoric-unit identities are stigmatizing and not gratifying; individuals will seek to build an identity around alternatives so that they create or sustain a favorable person-level identity while receiving positive reactions for playing out a more generalized role identity. For example, a person may consistently be aggressive in all interactions and may take great pride in this aggression and ability to "tell it like it is" in all situations. This person may constantly praise his or her honesty and candor and, in doing so, generally seem to feel pretty good about himself or herself, even while being quite irritating to any audience. This playing out of a more generalized role is often a strategy to avoid presenting other types of categoric- or corporate-unit identities or role identities from a stigmatized corporate unit. In downplaying these other possible alternative identities, and focusing on a generalized role identity where a person can present oneself in a favorable light (at least in the person's own mind), verification may be easier and less stigmatizing since most people do not want to question an aggressive person and, thus, by default allow the identity to play out. Humans can, through the activation of defense mechanisms, block out negative feedback from others that might give such aggressive individuals doubts about what they are doing. On the other side of this generalized role, people often put on a constantly happy, cheerful "front" (Goffman 1959) to hide real pain and depression over other identities, thereby allowing an aggressive person to feel some positive emotions by the reactions of others to such a happy-appearing person.

Humans are evolved from highly individualistic great apes and are supercharged with emotions and cognitive capacities to see themselves as an individual, standing in a world of others and social structures, where a self must be presented to others. This need-state, as part of the psychology complex, remains powerful even when it is difficult to verify an identity or to receive positive emotions from others over a stigmatized identity. Humans, then, are all doomed by their very nature to be vulnerable in any interaction where they can get one or more of their identities verified by others, particularly identities derived from categoric units, corporate units, and roles that are stigmatized by cultural beliefs and status beliefs. Presenting identities (identity-making) and reading gestures of others to determine their identities (identity-taking) are hardwired in humans' inherited legacy from the LCA of hominins and then expanded by the elaboration machine. Being so, they are part of human nature across all the complexes (emotions, cognitive, psychological, interaction, and community) of human nature. Because identity is tied to both emotional and motivational dynamics in other complexes, humans are inevitably driven to present their identities and to discover the identities of others during virtually all sustained interactions. Each person is, then, highly vulnerable to, and dependent on, others to verify self-presentations, with the result that interactions can often be on the edge of turning negative. Most of the time, however, they do not turn negative because humans have incredibly high acuity in reading the gestures of others and presenting gestures to others that facilitate accurate readings of not only identities but virtually every other dimension of interpersonal behavior. Such interpersonal acuity was clearly inherited by the ancestors of great apes and hominins, and therefore, could be selected on to generate even more acuity.

The Evolved Complexity of Role-Taking and "Theory of Mind"

Like great apes, humans engage in the reciprocal to role-taking, as George Herbert Mead argued in his lectures eventually published as *Mind, Self, and Society* (1934). Moreover, probably much more than great apes, humans engage in role-making, a concept originally proposed by Ralph H. Turner (1962). Thus, my effort above to convert the identity dynamics in interaction as a simultaneous process to identity-taking and identity-making simply extends conceptions developed by Mead and Ralph H. Turner. What I am doing in this extension is decomposing Mead's and Turner's ideas to emphasize that the dual processes of role-taking and role-making involve a series of more specific dynamics that should be conceptualized in more precise terms. Identity-taking and identity-making are one element of role-taking and role-making that should be separated from the notion of "role," although identities can form around particular roles. In this vein, I am proposing a more detailed set of processes that have been lumped for too long under the notion of role-taking as well as under the notion of a theory of mind proffered by primatologists, biologists, and evolutionary psychologists.

At the very least, it is in the nature of humans to engage simultaneously in the following processes during the course of an interaction (Turner 2011, 2019): (1) role-taking and role-making, (2) status-taking and status-making, (3) identity-taking and identity-making, (4) structure-taking and structure-making, (5) situational-taking and situational-making, (6) culture-taking and culture-making, and (7) emotion-taking and emotion-making. In the analysis of other human nature complexes, let me briefly elaborate on the more limited discussion in the previous chapter.

Role-Taking and Role-Making

Mead (1934) employed the label *role-taking* to include all the processes that addressed in this section. I include his conception here to take some of the conceptual burden off role-making (Turner 1962) as the master process and to downsize this concept to a more specific set of processes revolving around actual role dynamics. Individuals can use *status* as a guide to role-taking because almost every identifiable role is tied to status locations in the divisions of labor of corporate units (i.e., groups, organizations, communities) or to memberships in categoric units (e.g., gender, sex, ethnicity, age). There are, as noted earlier, some generalized roles, often related to emotions such as being happy, sad, angry, or depressed and to affective states such as being upbeat, serious, condescending, smart, and other culturally defined roles. Thus, I prefer to keep the notions of role-taking and role-making more delimited in order to conceptualize the many additional dimensions involved in the give and take of interpersonal behaviors. It is in the nature of humans, therefore, to seek to make a role for themselves (role-making) vis-à-vis others and at the same time to read others' gestures (role-taking) in order to determine the role that these others are asserting for themselves.

Status-Taking and Status-Making

Roles are often attached to status as being the appropriate behavioral component of incumbency in a status location within two types of units: (a) a location within division of labor of a *corporate unit* and (b) a membership in a *categoric unit* marking an individual as having what is often termed *diffuse status characteristics*, such as gender, ethnicity, class, and so on, that they carry with them in most situations. *Status-taking* and *status-making* seek to establish what status position in the division of labor of a corporate unit (group, organization, or community) is relevant, if any, and what diffuse status characteristics related to categoric units in society are relevant, if any. Thus, a person's gender is almost always relevant in human interaction, with somewhat different expectation states for males and females. This diffuse status characteristic can intersect with status positions in a corporate unit, such as a female manager or a male secretary. There will be normative expectations as well as more general cultural beliefs for positions in corporate units *and* expectations for diffuse status characteristics for categoric unit memberships; and thus, when individuals status-take, they decide which, if any, of these types of status are relevant and, if relevant, what the expectation states are for persons in these two basic types of status and what cultural beliefs or status beliefs are to be used in evaluating a given status.

Elaboration of social structures thus makes it important to understand the relevant structural units within which an encounter among individuals is embedded, thereby leading not only to role-taking but also status-taking in order to establish relevant expectations. The converse of this status-taking is status-making, where individuals seek to assert a relevant status or, alternatively, to play down status considerations associated with corporate or categoric units. As a result, humans can often implicitly negotiate which statuses are off the table for the duration of an interaction and which should be salient. Until individuals can agree on these matters, the interaction will be problematic, although most of the time individuals understand and agree on what status positions or memberships are relevant.

Because status-taking and status-making will orient individuals to roles, corporate units, and categoric units, this process inevitably flows into identity-taking and identity-making of the identities built around roles, corporate units, and categoric units. Hence, I will briefly review these identity dynamics that flow out of role-status-making and -taking. Similarly, role-status-identity-taking and -making also focus individuals' attention on the relevant structural units for an interaction, beginning with groups, organizations, and communities and then moving potentially to higher levels of social structure such as institutional domains, societies, and perhaps even systems of societies. Attention to structure also involves assessing diffuse status characteristics tied to the stratification system and ideologies legitimating inequalities inhering in this system.

Identity-Taking and Identity-Making

Flowing in concert with role and status processes in human interaction is, as emphasized earlier, a process whereby individuals seek to determine the salient identities being presented by others (identity-taking) and at the same time both consciously and unconsciously manipulate speech and body language to communicate to others the identity or identities that they are seeking to have verified by others (identity-making). Driven by need-states for positive emotions, individuals seek to have others verify in a positive way the identities that they present. Yet, since identities are often built around status locations in corporate units, status as membership in categoric units, and roles (whether those accompanying status or more generalized roles), all presentations of identities tied to status and roles are subject to evaluations by *status beliefs* in the culture of an institutional domain or stratification system. Thus, a person can have a stigmatized identity by its association with status beliefs and broader cultural beliefs that are negative, with verification simply reaffirming the stigma of a person that lessens the chances that this person will experience positive emotions. Individuals will generally seek to present only those identities that reveal more positive evaluations from status beliefs and more general cultural beliefs, which means that they will pick and choose which identities to bring forward. If an individual can bring forward several identities, which have varying degrees of stigma and valor, then often a less-valued identity can still be presented because a more highly valued identity dampens the stigma of the less-valued identity. A person in a stigmatized ethnic category (associated with prejudicial beliefs) can present this identity with some impunity if this person has another identity that is valorized by status beliefs, as would be the case if this person were president of an important corporate unit.

Humans make implicit strategic calculations, I believe, on just how to present identities. One strategy is to present only those that carry positive evaluations while repressing those that do not. Another strategy is to present intersecting identities that lead to a net positive evaluation, as in the example of the person with a devalued ethnic identity who is the president of a corporate unit. Yet another is to present only a role identity, disembodied from a structural unit and negative cultural values, such as being happy, upbeat, and otherwise constantly pleasant. Still another strategy is to emphasize the core or person-level identity, if this can be presented in a way that is positively valued by cultural beliefs. This person-level identity can be presented alone, as the sole identity, but more often, I suspect that individuals mix and match identities in order to get a positive overall outcome. Thus, people may invoke their core or person-level identity when needing to compensate for lower-valued identities that, given the circumstances, cannot be downplayed or ignored. Still another strategy is for individuals with a series of consolidated identities that are all stigmatized to limit interactions with those who hold these stigmatizing status beliefs and, instead, interact only with those who have similarly stigmatized identities. These kinds of interactions lessen the power of identity to affect emotions, with the interaction among individuals with a similar profile of identities producing the positive emotions necessary in all interactions (see later discussion of interaction rituals). Thus, gang members, homeless people, poor persons, members of ethnic minorities subject to prejudicial beliefs and discrimination, outcasts, and other stigmatized persons in general are wise when they interact with people like themselves. In this way, the positive emotions coming from interaction rituals, per se, perhaps supplemented by presentation of positive elements of a person-level identity, are enough for individuals to experience positive emotions even among those with stigmatized identities.

Structure-Taking and Structure-Making

Status-taking and -making are also part of what we might call *structure-taking* and *structure-making*, as individuals come to agreement on the relevance of which particular social structures humans have elaborated during societal evolution—groups, organizations, communities, institutional systems, societies, and intersocietal systems—are to be the reference point for the interaction. In fact, structure-taking and structure-making may precede role-taking/making or status-taking/making. These processes may all occur simultaneously. For simply organized hunter-gatherers, structural, role, and status processes were relatively easy to determine because there were only two corporate-unit structures (nuclear kinship units and band) and two categoric units (gender and age), creating a limited number of statuses and roles (mother, father, children) and a limited number of diffuse expectations (for males, females, and members of different age cohorts). Once humans began to elaborate new types of social structures, however, the number and diversity of corporate and categoric units increased, as did the number and diversity of institutional systems (e.g., kinship, economy, polity, religion, education, science, arts, law) inside societies and linking societies to each other. Thus, structure-taking and structure-making, along with status-taking and -making as well as role-taking and -making became much more complex processes, but with humans' expanded cognitive capacities negotiating these diverse elements guiding interactions comes rather easily.

The process is greatly simplified by most interactions occurring within a corporate unit within a community within an institutional domain within a society. This embedding of the layers of social structures built up from groups is easily understood because humans, like great apes and their ancestors, already look beyond the group to the community level of social organization, which is the most stable structure in great ape societies. Thus, with this aspect of structure implicitly understood, a good portion of the expectations in the interaction are automatically activated. If expectations associated with the categoric units in which an interaction seems embedded (e.g., the distribution of gender, ethnic, and other diffuse status characteristics), further fine-tuning of expectations can occur. As these are sorted out in further role-taking and -making, status-taking and -making, and structure-taking and -making, the relevant aspects of culture associated with the structure and the local situation within this structure can be picked up through culture-taking and culture-making.

Culture-Taking and Culture-Making

Accompanying role, status, identity, and structural determinations are efforts of individuals to mutually understand the elements of symbolic culture, particularly expectations for behaviors and demeanors, that are relevant and appropriate. At the same time, individuals are also trying to assert or "make" relevant and appropriate some elements of culture while communicating the less relevant and appropriate elements of culture. For example, an interaction in a business corporate unit involving only male employees at the same status who have been coworkers for a long time will invoke different elements of the broader societal and institutional culture to bear on the local expectation states associated with males, as a diffuse status characteristic, to the expectations associated with their particular location at the same place in the division of labor of the corporate unit. Consider a different interaction in the same corporate unit between a male and a female, with the female holding authority over the male, in the same sector of the economy. They will most likely invoke somewhat different cultural elements, particularly expectations for, and evaluations of, the behaviors of males and females (as categoric units) complicated by expectations for status and roles revealing different positions of power and authority (in the corporate unit), plus any other more general institutional norms. If we shuffle any of these elements—the nature of the corporate units are a small group of individuals embedded in a school corporate unit in a small or large community with few differences in authority but significant differences in categoric-unit memberships by gender, age, sexual preference in the institutional domain of education—then the dynamics of on-the-ground structure-taking and -making, role-taking and -making, identity-taking and -making, and status-taking and -making become much more complicated. Yet, with their elaborated cognitive and interaction complexes, humans will normally be able to interact without too much stress. They will be able to access their stocks of cultural knowledge on the basis of their culture-taking on the basis of their assessing others' status-, role-, structure-, identity-, and culture-making efforts. This process will normally proceed smoothly and rapidly as adjustments for varying expectation states on, and evaluations of, individuals in different categoric units are made and, to add another complication, as some individuals seek to assert or make their statuses, roles, identities, and cultural expectations that are somewhat unique. Such is the power and flexibility of humans' evolved nature.

Situation-Taking and Situation-Making

As individuals engage in these processes just described, they are also fine-tuning their actions on the basis of the particular features of the situation in which interaction is occurring. Initially, individuals will seek to categorize (a) the nature of the situation and (b) the nature of persons, somewhat along the dimensions summarized in Table 10.1 that cross-tabulates the nature of situation as *work-practical*, *social*, or *ceremonial* (Collins 1975) against the nature of others in the situation as *personages*, *persons*, and *intimates*. Humans store information in these kinds of categories, and as they enter situations, they check on the ecology of the situation, the demography of who is co-present, and the status of those present. At the same time, individuals cognitively search their stocks of knowledge for the appropriate location delineated in Table 10.1 with respect to (a) the nature of persons who are likely to be encountered and (b) the nature of the basic situation, which can be discrete in terms of the categories across the table but, more likely, some mix of the three. Moreover, the person co-present may fall into all three categories, thereby requiring constant interpersonal adjustments as a person presents self to intimates, persons, or personages. Most of the time, humans have little difficulty in making adjustments as they move from one person to another.

The nature of interaction changes with shifts in situational parameters along dimensions outlined in Table 10.1 but also along such dimensions as how enclosed is the ecology, how open is movement through space by others not involved in the interaction, how dense is the space and juxtaposition of bodies, what are the props and other physical properties of the space, and so on. All these varying elements of situations carry cultural meanings and shift expectations for how people should behave. At times, when a situation, such as a line at a movie theater or an elevator at a shopping mall, brings together two or more individuals

TABLE 10.1 Initial Categorizations of Others and Situations

Nature of others	Nature of Situation		
	Work/Practical	*Social*	*Ceremonial*
Personages	Others are functionaries, even strangers, for achieving specific task goals and, thereby, treated by their roles in coordinated activity.	Others are of larger ongoing social situation and are owed appropriate stylized responses fitting the social situation	Others are strangers toward whom informal, polite, and response responses are owed
Persons	Others are functionaries in coordinated activities are to be treated as unique individuals in their own right	Others are owed informal response to the collective social occasion, as well as responses reinforcing their self presentations as unique individuals.	Others are owed polite, somewhat informal responses appropriate to the ritual activities being conducted.
Intimates	Others are close friends whose actions are relevant to achieving goals through coordinated efforts, but whom are owned diffuse obligations and emotional responses	Others are to be treated as close friends, no matter the nature of the social situation, to whom diffuse obligations and emotional responses are owed.	Others are to be treated as close friends, regardless of the nature of the ritual occasion

with different degrees of authority in the same corporate unit, such as workplace, the interaction can become awkward and will generally be highly ritualized if the expectations of how to behave are ambiguous over how much of the workplace culture, status, roles, and structure should be invoked in what is a more public and informal setting. Yet, humans with their inherited interpersonal skills, elaborated and expanded by natural selection as well as by the other evolved complexes of human nature, can generally muddle through such awkward moments.

Emotion-Taking and Emotion-Making

Every dimension of interaction is laced with emotions. And so, as individuals role-take and -make, status-take and -make, identity-take and -make, structure-take and -make, culture-take and -make, and situation-take and -make, a great deal of *emotion-taking and emotion-making* is also occurring as individuals seek to determine what emotions, emotional demeanors, and elements of emotion culture of the structural unit within which an interaction is occurring are relevant and appropriate. Each individual can also emotion-make by "putting on" a sad, happy, angry, worried, or some other "face" to manipulate others' emotion-taking; and these others will by their counter emotion-making communicate if this is acceptable in a particular situation. Emotion-taking and emotion-making are perhaps the one force that can disrupt and breach interactions if individuals do not reach consensus over what are the appropriate emotions to be felt and, more importantly, to be expressed, especially if the emotions are associated with success or failure for identity-taking and -making. Indeed, establishing the appropriate emotional "mood" is an important part of keeping an interaction on course.

Interactions are often difficult because they are complex, but the real danger to an interaction is the failure to get the emotional tone right. Thus, it is in humans' evolved nature to try to establish an emotional tone or collective mood to the interaction because when emotions clash, every other aspect of the interaction is also breached, forcing individuals to engage in painful repair work. In almost all interactions among humans, they engage in a preliminary emotion-taking almost immediately in order to get a sense for what the emotion-making gestures emitted by others mean for all other forms of "taking" and "making" in an interaction. In making humans highly emotional, natural selection created both the basis for increasing social ties and solidarities, on the one side, and the basis for disrupting social ties and solidarity, on the other. Thus, duality is simply part of humans' evolved nature.

Theory of mind as conceptualized within biological anthropology and primatology (Premack and Woodruff 1978) and more generally within biology and neurology (Hare, Call, and Tomasello 2001; Povinelli and Vonk 2003; Povinelli, Nelson, and Boysen 1990: Mitchell 2011a, 2011b: Meltzoff 2002; Parr et al. 2005) *glosses over* the more complicated processes that I have just outlined. This conception is basically the same as George Herbert Mead's early-twentieth-century conceptualization of *role-taking*, revolving around the capacity and propensity of humans to read the internal mental states of others and their disposition to act in certain ways in order to cooperate with them. However, the preceding review of the many dimensions of interaction is meant to illustrate that theory of mind is far too simple when applied to humans, and still too simple when applied to great apes as well as other highly intelligent and emotional mammals and birds.

For humans, what Mead termed *role-taking* is a much more robust process than that among the great apes because of the evolution of the various complexes examined in earlier chapters, but humans' ability to read gestures to interpret others' internal mental states, coupled with their propensity to manipulate presentations of self to others, are nonetheless only extensions of what great apes can now do and, hence, by the logic of cladistic analysis, what humans' hominin ancestors could do. As natural selection grew the subcortex and neocortex of hominins, which, in turn, eventually allowed for speech and cultural production, the nature of interpersonal dynamics became much more complicated but still fundamentally the same as evident in the other descendants of the LCAs, present-day great apes.

Framing and Interaction

Erving Goffman Simplified

As outlined in Chapter 8 on the cognitive complex, framing is an organizing process whereby the brain orders information in a process analogous to the way in which frames are used by computers to order knowledge for rapid retrieval. About the time that "frames" and "framing" were being developed in computer science, sociologist Erving Goffman (1974) was developing a conception of frames and framing to understand interaction as a key phenomenological process. Like counterparts in computer science, Goffman emphasized that the brain generates frames that delimit the range of information stored and retrieved in interpersonal behaviors. Goffman's vision of frames and framing was, as I emphasized earlier, more complicated than it had to be; and in fact, it was probably too complicated to achieve what frames can do: denote what is to be part of the interaction and what is outside the "frame" as excluded from the range of phenomena.

Keying and Rekeying of Frames

Given the complexity of human interaction, it is not be surprising that natural selection hit on the same solution as computer scientists: order and retrieve information needed in an interaction through signaling by individuals (in talk and body language) of the relevant frames—what is to be part of the interaction and what is not to be part of the interaction. It is also unsurprising that there would be interpersonal practices that, in Goffman's terms, *key* the frame—if frames are to be changed, then there would be interpersonal practices for *rekeying* the frame that guides an interaction. Goffman's discussion of *primary* and *secondary frames* and the process of *lamination* of frames was, I think, too complicated to be actually employed in moment-by-moment interaction, but the more general insights that humans key and rekey frames to keep everybody "on the same page" of cultural scripts is nonetheless fundamental to human interaction.

Keying and rekeying appear highly ritualized in talk and body language because rituals call attention to shifts in the interpersonal flow and activate emotions if these shifts are not honored by others. To take an obvious example, people often say these days "let's not talk about politics" as a ritualized verbal statement to rekey the frame guiding an interaction. If this appeal for rekeying is not honored by others, negative emotions will be immediately aroused. If honored, the person proposing the rekeying and the others with whom this person is interacting can scan their *stocks of knowledge at hand*, to use Alfred Shutz's (1932) image, to retrieve the relevant stores of knowledge to keep all participants in the interaction on the same *footing*, to use Goffman's term.

Frame-Taking and Frame-Making

Figure 10.1 outlines basic axes of framing during human interaction (Turner 1988, 2002, 2007, 2010). Individuals store information about appropriate behaviors with respect to the bodies of others and the degree to which personal information is to be part of an interaction, as is noted at the top and bottom frames delineated in Figure 10.1. With respect to *body frames*, such fundamental issues as the acceptable distance between,

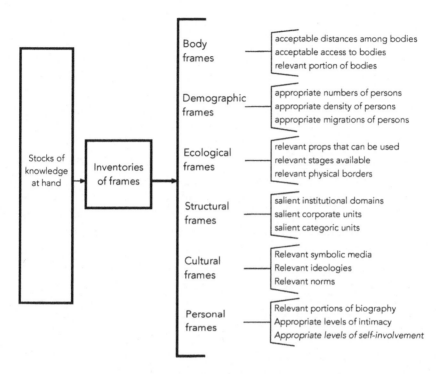

Figure 10.1 Cognitive Framing as an Ordering Mechanism

degree of access to, and relevant positions of bodies are always part of a keying process, typically depending on other framing axes, such as cultural or organizational frames. There are also often *demographic frames* that set the appropriate number of persons that should be present, the tolerable density among them, and the degree to which migration into and out of the interaction situation is appropriate or even allowed. Physical frames about what *props*, what *stages*, and what physical boundaries and *borders* can be used in an interaction are always present. All these various classes of frames, involving specific ritualized interpersonal behaviors and forms of talk to invoke, are highly constrained by the structure of the situation as marked by *organizational frames* with respect to what institutional spheres are relevant (e.g., kinship, religion, economy), what corporate units within institutional spheres (e.g., family units, churches, workplaces), and what categoric units (e.g., age, sex, gender, class, religion, ethnicity/race) are relevant in the interaction. Associated with organizational frames are also cultural frames about which values, beliefs/ideologies, and norms are to be relevant.

The process of framing is, first of all, simplified by the status and roles that people are playing in a given situation within a corporate unit within institutional domains. Once this information on structural location is noted, then frames with respect to culture, bodies, persons, demography, and physical properties of the situation are more readily invoked. Such is always the case for how knowledgeability about frames is stored in the brain, with individuals developing sets of syndromes of related frames that can be rapidly brought to, and understood by, self and others in a situation. As part of humans' storage of knowledge, we all implicitly "know" the relationship among frames for bodies of various categories of person and for how biography, intimacy, and self-involvement are to be framed in various types of demographic situations with particular props, stages, and borders within a particular organizational context and the culture attached to this context. For example, people have little trouble distinguishing between a workplace and the office picnic, although problems can emerge with the effects of drinking alcohol at the office party. Still, most persons can immediately shift to a syndrome of frames listed in Figure 10.1 rather effortlessly and unconsciously. When they know the structural location and relevant organizational frames, other frames will easily fall into place. If frames are shifted to something that is out of the ordinary for a particular type of situation, the rekeying will typically be highly ritualized and visible to all persons.

These dynamics are more critical in complex societies where there is a larger number of persons organized in a wider variety of differentiated structures and systems of cultural expectations, but these dynamics were operative even in simple societies. The ability to make frames and to alter or rekey them allowed for much more of the complexity of this process to be managed, typically rather easily by most persons in most situations. Thus, it is likely that the basic propensity of humans to frame all situations allowed for human populations to build up not only their numbers but also the complexity of structures and cultural systems organizing these larger populations. Just as computers are made bigger, faster, and smarter by programs that frame information, so humans are smarter and more interpersonally adaptive by their neurology that, it seems, "naturally" frames situations so as to keep interactions focused, while at the same time providing a mechanism—ritualized and episodic keying and rekeying of frames—to keep interactions on track. Only in situations where framing is not clear to individuals—such as a lower-level worker meeting a high-level superior outside of the normal context of the workplace—does framing often become difficult, although most often the default frames of how to interact in polite ways are invoked as a means to get the framing process started. These general frames include somewhat modified frames of interacting with superordinates coupled with clear frames for low access to bodies and personal knowledge and frames about how to migrate away from authority or subordinates in polite ways utilizing props, borders, and stages. We all implicitly know how to deal with this "awkwardness" without breaking the implicit stack of frames within which this interaction must take place and, typically, disband in face-saving ways for all parties. Without the ability to assemble not just one but several frames simultaneously (Goffman's notion of "lamination"), such interactions would be difficult. Indeed, societal complexity could never have evolved if humans did not have as part of their "nature" this rather amazing capacity to frame and reframe interpersonal interactions.

I have come to see framing and rekeying of frames like many other interpersonal processes: as a constant effort to read the frames being implicitly suggested by others through what we might call *frame-taking*. It is in humans' nature to constantly assess which and what kinds of frames, along the lines outlined in Figure 10.1, are being asserted or presented by others because having this knowledge makes all other interpersonal processes much simpler. We know the boundaries of expectations for what the other person wishes to occur, thereby generally making it easier to role-take and -make, status-take and -make, identity-take and -make, structure-take and -make, situation-take and -make, culture-take and -make (especially since frames are a part of culture and are often involved in the storage of cultural codes to be retrieved rapidly by the prefrontal cortex), and emotion-take and -make. To complete the notion of frame-taking and frame-making, all humans are also often engaged in frame-making, by which they often emit ritualized phrases to mark an effort to rekey a frame. These frame-taking and -making processes can occur simultaneously or in phases as an interaction proceeds, but it is the capacity to do both that allows humans in complex societies to keep interactions on track.

Language, Rhythmic Synchronization, Ritualizing, Totemizing, and Exchange

Language and Rhythmic Synchronization

Human interaction operates along several media of communication: (1) the language of emotions (as revealed by facial, eye, and body gesturing) and

(2) speech acts revolving around verbal turn-taking often infused with affective states and intonations. As is the case with great apes, the beginning of any human interaction is initiated with ritualized greeting revolving around stereotyped sequences of communication through vocal calls and body language, with humans adding fine-tuned talking rituals and body rituals also resembling a gestural language carrying common meanings. Greeting rituals are thus critical to getting an interaction off on the right footing and perhaps also inside the right frame, which, it is implicitly presumed by its participants, will lead to a positive emotional experience.

Communication is almost always laced with emotional content and is, thereby, part of the process of emotion-taking and -making. Of particular importance for humans is the degree to which (1) ritualized greetings (and, later, departures), (2) turn-taking in speech acts (Sachs et al. 1974), and (3) emotional cues expressed in face, eyes, bodies, and inflections of voice will all fall into a *rhythmic synchronization* (Collins 2004), because this kind of synchronization leads to the arousal of positive emotions that are essential to stronger social ties and group formations revealing collective solidarity. When human interactions are "out of sync" along these three dimensions, mild negative emotions disrupt the smooth flow of interaction. To avoid being out of sync, it is thus necessary to have ritualized exchanges of greetings. These greeting must be followed by verbal turn-taking in speech acts accompanied by emotional expressiveness. If these speech and emotion-arousal processes promote rhythmic synchronization of talk and bodies, the interaction can proceed and is likely to generate positive emotions among its participants. Indeed, these early phases of what Randall Collins (1975, 2004) terms *interaction rituals* can often escalate the arousal and exchange of positive emotions in ways that add another valued resource to the exchange (positive emotions) and allow individuals to meet need-states for profitable exchange and for experiencing positive emotional arousal. The evolved nature of humans is to engage in such rhythmically synchronized interactions because these were critical to the enhancing of emotions essential to forming the stronger social ties and group solidarities that enabled hominins and then humans to survive. This propensity is clearly evolved from the "carnival" behaviors of chimpanzees where animated behaviors fall in synchronization and arouse positive emotions, apparently symbolizing the solidarity of the chimp community. Humans have simply amplified this propensity and extended it to each and every interaction. Indeed, the dependence of humans on rhythmic synchronization is so great that when it does not occur, interaction is strained and often arouses negative emotions.

Shorter-Term Rituals

Most of the rituals that allow individuals to initially fall into rhythmic synchronization are what I term *shorter-term rituals*, like those used in greetings of others and closing off an interaction, or as is the case when parties to an interaction seek to impose an initial frame or rekey the frame during the interaction. These rituals involve short bursts of stereotypical speech, and, if needed, associated proper body language, to propose an altered path of interaction. Thus, an enthusiastic greeting ritual accompanied by a hug and kisses sets a frame for the interaction that is very different than just, "oh, hello, how are you." The rituals are short, but they often establish the flow of the interaction for the duration of the encounter and, perhaps, even set up expectations for next time the interaction is taken up. Indeed, the short closing ritual often determines just how the parties in an exchange will greet each other in the next encounter. If the closing ritual is unenthusiastic, the greeting rituals initiating the next interaction will be tentative and halting. So, short rituals have large effects in opening an interaction, and these rituals sustain a chain of interaction rituals over time (Collins 2004).

Longer-Term Interaction Rituals

The term *interaction rituals* was originally coined by Goffman (1967) and, somewhat later, by Collins (1975, 2004) as *interaction ritual chains* to describe more protracted solidarity-generating dynamics. What these theorists essentially did is "downsize" the big collective carnivals of chimpanzees and humans to a quieter but probably more important bonding mechanism in which the assembling of humans in propinquity almost automatically generates a sense of co-presence and a propensity to talk by use of short-term rituals, often ritualized greetings that, if they continue, move into sequences of talk and displays of emotions, voice inflections, and tones as well as body language that become rhythmically synchronized (Collins 2004). This rhythmic synchronization arouses additional emotions and then, if the interaction is sufficiently emotional and/or repeated over time, it is symbolized in some manner, with references to the symbols marking and denoting collective emotions promoting solidarity. Goffman tended to see rituals as mostly short term, whereas Collins sought to extend the notion of ritual to a more prolonged series of actions, or chains, across encounters (Collins 2004). Figure 10.2 outlines the model proposed by Collins (2004), albeit in a somewhat altered form from his presentation.

As is evident, this model of human interaction rituals delineates a behavioral propensity clearly inherited from the LCAs of chimpanzees

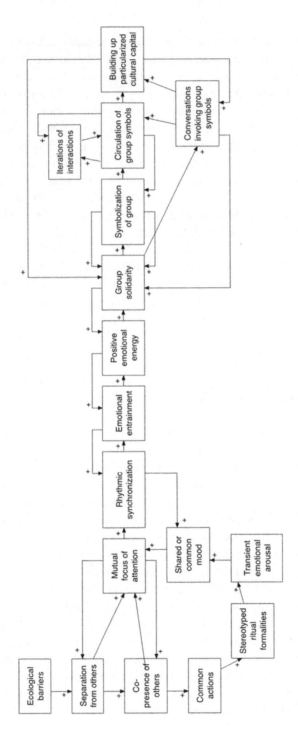

Figure 10.2 Collins' Analysis of Large-Scale Interaction Rituals

and hominins because it employs the mechanism of emotional arousal evident in chimpanzee "carnivals." The ritual is, however, more complex because humans can talk and, thereby, more readily fall into rhythmic synchronization during conversational turn-taking (Sachs et al. 1974). Of course, humans have a larger palette of emotional states to work with; and most importantly, humans have culture that can enshrine and symbolize relationships in normative expectations and totems marking the relationship as something special. As hominins' emotional repertoire expanded and, then, as the brain began to grow, nonverbal gesturing through emotions was increasingly supplemented by talk and speech. Talk and speech allowed for the codification of culture, so the "carnival" of the LCA of chimpanzees and hominins could be transformed into more protracted interaction rituals that can occur almost any time that humans interact. Over time, as interactions were iterated, they became symbolized by totems and normative prescriptions and perhaps proscriptions. Emotions are constantly flowing back and forth during these rituals, and this circulation of emotions gives the symbols representing relationships among individuals the power to generate stronger social bonds and solidarities among humans. For, despite their big brains, spoken language, and culture, humans still retain some of the weak-tie propensities of their ancestors, with the result being that individuals must "work at" interactions and use emotions to form attachments and solidarities with each other. Thus, it is the *interaction ritual* delineated in Figure 10.2 (Collins 2004) that orders the flow of interaction leading to heightened positive emotions, to increased social bonds, and to cultural symbolization of relationships in totems and normative expectations. Refinements evident in human interaction rituals are built on the neurological base evident in great apes, especially chimpanzees. It is not a far step from the chimpanzee carnival to the friendly interaction among humans finding themselves in co-presence, where a sense of perhaps only temporary solidarity is built up during the interaction ritual, thereby creating positive emotions and thus commitments to the larger social structures within which interactions occur (for more details, see Collins 2004; Turner 2002, 2007).

Totemizing

Interaction among humans is much more likely to be "totemized" by physical objects or symbols than that among great apes. The capacity to generate culture leads humans to verbally designate and thus symbolize this culture, often with physical totems—monuments, uniforms, flags, artifacts—as well as with verbal labels such as titles and phrases of speech. Thus, virtually any object or word/phrase that denotes *the important properties of an interaction* within various kinds of social structures can symbolize a collectively organized unit—from a group through organization and community to an entire society. Indeed, interaction with a priest in appropriate clothing, with a member of Hells Angels in their "uniforms" and customized motorcycles, with members of a church congregation in a building with a cross on it, even a meeting of friends at a particular booth in a bar, and a flag symbolizing a nation can all denote social solidarity and activate the emotions generating their solidarity. When ritualized actions and verbalizations directed at these symbols of solidarity are executed, these behaviors arouse the emotions behind this solidarity. Speech, writing, and capacities to forge cultural meanings allow humans to "totemize" and then direct emotion-arousing rituals toward these totems and the underlying reality that they symbolize: social solidarity (Durkheim 1912). Great apes show the beginning of this propensity, but it took the evolution of the elaboration machine to make such totemic behaviors widespread as mechanisms for solidarity-producing interactions in virtually any gathering of humans.

Indeed, as Emile Durkheim (1912) came to believe after 1895, the central mechanism by which societies are integrated is the capacity of humans to create totems symbolizing their group-level solidarities. Totems not only mark but also arouse emotions of solidarity and collectively direct humans to the totem that symbolizes their solidarity with each other; this capacity can be extended, Durkheim argued, to larger social structures where individuals are not engaged in direct interaction. When diverse subpopulations of a society accept and legitimize a common totem, such as the flag symbolizing a whole society, and are able to engage in emotion-arousing rituals toward this common symbol, then diverse and differentially located members of a society can create collective solidarity to the "collective representations," or systems of cultural beliefs, ideologies, prescriptions, and proscriptions, symbolized by the totem. Thus, totemizing began in smaller group formations, but the cognitive and emotional capacities of humans, coupled with their elaborated interpersonal skills, enables them to "worship" common symbols even if the individuals engaging in such worship never interact. This power of commonly held beliefs or "collective representations," to use Durkheim's term, that have been totemized in physical and linguistic symbols dramatically increases the capacity to build larger-scale social systems, considerably beyond the groups and bands of hunter-gatherers. Elaboration of emotional and cognitive capacities, coupled with language and the ability to generate collectively held beliefs (e.g., *collective representations*), allowed for increased solidarity among members of very small populations and created the critical preadaptation for the evolution of mega societies by providing a fundamental basis for the integration of large, diverse populations.

Exchange

Human interactions, like those among great apes, almost always involve implicit or explicit *exchanges of resources*, but among humans, multiple resources are often in play, above and beyond extrinsic resources. Humans determine the degree to which self and identity, status, respect and dignity, and conformity to cultural norms are being realized, and to the degree that they are, humans experience positive emotions. Conversely, when self, status, dignity, and conformity to cultural expectations are not realized, negative emotions arise. Like interactions among great apes, positive emotions are always being exchanged in human interactions, but there are many more dimensions along which exchanges facilitate activation of all those processes arousing positive emotions—from the verification of identities through successful "taking and making" along the lines enumerated earlier to both spoken and emotional languages falling into rhythmic synchronization. Along many other dimensions exchanges can go astray and thereby arouse negative emotions. The positive emotions aroused during the course of exchange become another highly valued resource because they are the resource that affirms self and that leads to smooth interactions, stronger social ties, and group solidarities sustaining human societies. A much more critical resource to human well-being and survival than the explicit and/or extrinsic resource being exchanged, then, is on the line in human exchanges: positive emotions associated with "fair" exchanges generating a sense of "profit" of costs and investments necessary to receive resources that may also be involved in affirming identities. These emotions become yet another additional and highly valued resource to the exchange; and indeed, positive emotional arousal is the resource that generates social solidarity.

The need for humans to experience justice in the resources given and received in exchanges was hardwired in great apes and, hence, hominin. If the positive emotions arise with greeting rituals, verification of identities, rhythmic synchronization, and successful "taking and making" along the six dimensions outlined , this conflation of emotional valences tied to the exchanges of *any* resources makes exchanges a powerful force of human nature. Because exchanges and the positive emotions that they can generate were critical to the solidarities that allowed hominins and then humans to organize and, thereby, survive in open-country habitats, these exchange dynamics became an even more essential trait of human nature.

Much like totemization as a preadaption to macro-level patterns of social organization that would allow large numbers of diverse and spatially scattered persons to remain integrated, the power of exchange had this same potential. Exchanges among preliterate populations enhanced solidarities among exchange partners within and between populations and subpopulations in smaller-scale societies. The hardwired propensity to seek fair exchanges yielding an excess of resources over the investments and costs involved in securing these resources was critical to increasing attachments, stronger ties, and group solidarities among hominins and early humans. It was the institutionalization of exchange processes, in differentiated markets of developing societies from late horticultural phases of societal evolution to the present, however, that allowed for the evolution of larger and more complex societies because of the inherent capacity of exchanges to promote solidarity.

Humans are thus wired to seek fair exchanges of resources yielding a profit of resources. As ever more aspects of society are mediated by market exchanges in virtually all institutional spheres, profitable exchanges arouse positive emotions and, thereby, enhance commitments to diverse levels of social organization if they yield perceived profits and are seen as fair. Workers taking wages for work, students paying for education and credentials, couples marrying for love but also the resources that they bring to a family, members of churches paying fees to receive the guidance of supernatural powers, persons paying prices for tickets to receive emotion-arousing entertainment, shoppers buying products (from food to almost anything) that bring rewards, taxpayers giving money to governments for valued services, and so on for any of the many exchanges that occur within and between institutional systems in complex societies generate positive emotions if seen as fair and profitable. From this constant flow of positive emotions, humans develop attachments not only to the exchange partners but to the larger exchange systems organizing institutional activity in a society. This inherited capacity and propensity to assess fairness and profits in exchanges became, in turn, one of the bases for humans' capacity to moralize social relations, once the capacity for speech allowed for the articulation of moral codes.

As Georg Simmel (1906) argued, industrial societies generate many more potential opportunities and avenues for receiving value that provide the potential for generating constant flows of positive emotions from receiving what is perceived of value. From these positive emotions, moral commitments to larger-scale social structures could be generated at the same time that micro-level positive emotions were generated with each and every localized exchange. Today, the constant flow of products in markets with giant players like Amazon take these dynamics to a new level. Consumers constantly experience a sense of value that keeps them scrolling in search of new products. This behavior, perhaps somewhat subversively, generates larger-scale commitments to the institutional system of economy and the broader society in which corporate interest co-opt individuals. As with cell phones and social media, commercial transaction mediated by media have addictive qualities, but they also generate diffuse commitments that can be hugely legitimating in macro societies. Yet, all these forces are only extensions of something evident in chimpanzees and hence the LCAs of chimpanzees and hominins and, as I will address in Chapter 12, also create severe problems for an evolved great ape.

Conclusion: The Evolved Interaction Complex

The *interaction complex* is, as the name suggests, rather complex. Human interaction is complex because, in essence, the viability of society is at stake, as is the well-being of each individual in interaction, during the course of each and every interaction. Emotionally charged-up interactions were the basis of hominin survival, as selection pushed for stronger ties and increased solidarity. Many need-states from the *psychological complex* must be met during interaction, and humans' emotional and cognitive complexes allow for use of many different emotional and cognitive hooks and strategies to forge social bonds and to sustain group-level and societal-level solidarities. It is the viability of humans' expanded patterns of social organization as made possible by the *community complex* that is on the line.

Perhaps humans' most remarkable capacity is the seeming ease with which complex interactions can be created and sustained. Yet, even as things are going well in an interaction, the potential for disruption if negative emotions are aroused is always present. Humans have to work hard at keeping group-level interactions going, and even though most of us can interact with some ease, there is always the potential for a breach in the interaction and the resulting arousal of negative emotions that are often difficult to put back in the bottle. It is in our nature as humans to be constantly attuned to what can go wrong, even if it is not at the forefront of consciousness, because in the simple societies of humans' origins, disintegration of a group could mean real difficulties in surviving in predator-ridden and often ecologically difficult habitats. The fragility of humans' emotions on such artificial and mediated groupings as "friends" on Facebook attests to how vulnerable humans are. Thus, along with the evolved emotions complex, the evolution of a more dynamic interaction complex was the key to human survival, but these two complexes are points of vulnerability as well. This complex is outlined in Figure 10.3.

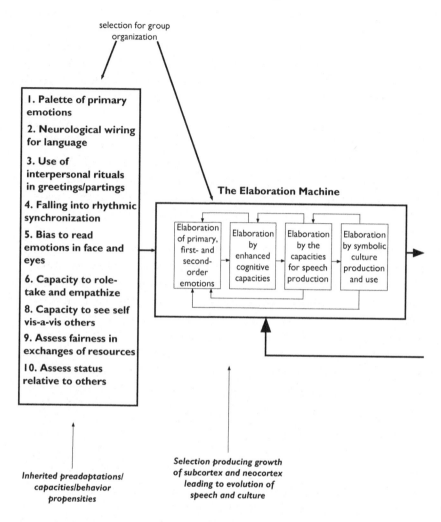

Figure 10.3 The Evolved Interaction Complex and Human Nature

1. All complexes of human nature can only be activated by exposure of the very young to interaction with other humans, which is facilitated at birth by infants' biological drives to seek out interaction and physical contact with humans.

2. By virtue of interaction with others and by emotionally responding to the gestures of others, humans develop a series of identities—minimally, *person-level identities*, *categoric-unit identities*, *corporate-unit identities*, and *role identities*—that they then seek to verify in virtually all interactions in order to have positive feelings about self.

3. In all interactions to varying degrees and extents, humans read the gestures of others and the context of the interaction to determine the (a) *identities* being presented for verification by others, (b) the *roles* that others are trying to play, (c) the *status* locations and memberships that others seek to occupy, (d) the *structure* that others seek as relevant to the interaction, (e) the *situational features* within these structures that others see as the most relevant and important, (f) the *cultural symbols*, texts, totems, codes, and frames that others see as guiding the flow of interaction, (g) the *emotions* that others are experiencing and that others see as appropriate, and (h) the *frames* guiding the flow of interaction. These efforts to "take on" the perspective of others in an interaction can be labelled *identity-taking, role-taking, status-taking, structure-taking, situation-taking, culture-taking, emotion-taking,* and *frame-taking.*

4. In all interactions to varying degrees and extents, humans always seek to "make" for themselves by presentation to others of both conscious and unconscious gestures communicating (a) one or more of their *identities*, (b) *roles* that they seek to play, (c) *status* locations and memberships that they want others to see as relevant, (d) *structures* that they see as relevant, (e) *situational features* that they see as relevant, (f) *cultural symbols*, texts, totems, codes, and frames that they as appropriate for the interaction, and (g) *emotions* that they are experiencing as well as emotions that they see as appropriate. These "making" processes of

self-presentations can be labelled *identity-making, role-making, status-making, structure-making, situation-making, culture-making, emotion-making,* and *frame-making.*

5. Human nature is thus driven and constrained by a continuous, mutual effort to exchange both extrinsic and intrinsic resources with others in which individuals receive more resources than the costs incurred in receiving them, with successful exchanges creating a new resource—positive emotions—that is added to whatever else in being exchanged, thereby increasing positive emotions and commitments to the exchange.

6. Humans are motivated to engage in both shorter-term and protracted ritual activities to create and sustain interactions promoting the exchange of positive emotions while enhancing social ties and collective solidarities. (a) *Short-term* rituals revolve around stereotypical behaviors that open, close, and structure the interaction among individuals during greetings and closings of interaction and at strategic points while in engaging in points 1–5 and 7, which follows. (b) *Longer-term interaction* rituals revolve around building on short-term rituals to raise the level of positive emotional flow, to increase solidarities, and to sustain the positive emotional flow and solidarity in iterated encounters strung together over time through fair exchanges of resources leading to the exchange of positive emotions that, in turn, lead to totemizing the interaction with symbols toward which emotion-arousing short-term rituals are enacted to sustain collective solidarity over time.

7. Humans in all interactions are motivated to experience positive emotions about self through the activation of the dynamics outlined in points 1–6; and as positive emotions are experienced, individuals develop commitments to others and the structures and cultures within which an interaction occurs.

11

The Evolved Community Complex and Human Nature

Community as the Basic Organizational Units of Great Apes and Hominins

The last common ancestors (LCAs) of great apes and hominins evidenced a unique pattern of social organization when compared to most mammals who are kin- and group-oriented. Great apes evolved in the terminal areas of the forest canopy and thus could not form permanent groups, as noted in Chapter 2 and elsewhere. The only stable unit of organization was a sense of community or home range that could be quite large, as much as 15–30 square miles, as is the case with come common chimpanzees today. Even as some great apes and hominins began to live in more terrestrial habitats and to form limited group ties, the more inclusive community was still the structural backbone of each population. With their comparatively large neocortex, great apes and hominins could cognitively map the ecological boundaries of the community as well as the demography of who belonged and who did not belong in the community. Given that these communities could be many square miles, members would wander around their home range and not see each other for weeks, even months, and yet could engage in greeting rituals when meeting up and proceed to role-take with each other and, it can be presumed, see themselves from the perspective of others. Unlike most other mammals, this kind of community orientation was sustained by weak social ties and a lack of kinship beyond mother and young children (ties that were broken at puberty as adolescents transferred to other communities or different locales in the same community; fathers were, of course, unknown because of promiscuity). The interaction complex evolved from the rather sophisticated interpersonal skills that great apes and hominins possessed. These same skills of being able to know all community members, even if not seen for a time, and to interact with them sustained this structure and its ecological boundaries.

The interaction complex and community complex are related in that they allow humans to interact with relative "strangers" and to sustain weak ties across larger territories—basic conditions necessary for mega societies of many millions of inhabitants. Only the social insects, such as ants and termites, can build such large societies revealing divisions of labor. Most mammals cannot create such societies because of their local group orientations as well as because of kin selection, or a preference to stay with and support closely related kin. Humans can, as Moffet (2018) provocatively proclaims, create "swarms" among very large animals, or, as Peter Turchin (2016) has termed it, can create and live in "ultra societies."

Inherited Traits and Effects of the "Elaboration Machine"

The inherited traits of LCAs subject to selection and then enhancement by the elaboration machine during late hominin and early human evolution are listed on the left side of Figure 11.1 on page 250. Subject to the powerful forces of the elaboration machine, the capacities to reckon community, to "know" its boundaries and its demography of inhabitants, to see self in relation to community, to engage in relaxed weak-tie interactions with community members, to be mobile and sometimes alone in movements through the community, to be able to mobilize emotions in quasi celebrations of the community when larger numbers of its members came into co-presence, and to defend the boundaries of the community with temporary groupings in patrols along the community's boundaries became the basis for humans' capacity to form *new layers of social structure* and, eventually, their cultures, which would lead to the large industrial and postindustrial societies of today. Early on, the sense of community among hominins could have been expanded to the territory of the nomadic hunting and gathering band. There may have even been a larger territory for all bands sharing the same language and cultural traditions, thus extending the territories to be conceptualized by many more square miles and requiring new kinds of more "diplomatic" interaction with members of other bands in this extended "home range" of sets of bands.

When humans finally began to periodically settle down from their more nomadic hunting and gathering, perhaps for only part of the year, these territories may have been defended, as is the case with chimpanzees today. By 10,000 years ago, as the human population began to grow, early Big Man systems of hunter-gatherers settled near sources of water, and fish began to be supplemented by horticultural communities (gardening without the plow or animal power) that were linked together because of unilineal kinship ties that began to be superimposed over communities. Similarly, pastoralists who were more mobile also began to use unilineal kinship as a means to organize nuclear families across ever more extended territories and to organize trade relations among subpopulations in a territory. As more advanced horticultural societies evolved, some were organized around communities connected to a dominant larger city housing political and religious elites that were beginning to operate as an early "state" (e.g., much of Central and South America before European conquest). Others were based on more extended trade networks among cities and more rural areas, typically accompanied by political domination by a "capital city" and its control over larger territories or regions (as was the case in China and parts of Europe and Asia before the use of the plow). As full-blown agrarian societies evolved (using the plow), often generating a central capital (e.g., Rome) and a large territory of conquered peoples (e.g., the Roman Empire) and, later, patterns of feudalism and smaller state-like structures (the "dark ages" after the fall of Rome), the community system had greatly expanded or, alternatively, was supplemented by many new layers of social structure: capital cities and various kinds of specialized rural and urban formations, such as cities that were devoted to trade and market activities; new institutional systems increasingly organized into bureaucratic-like structures in highly differentiated political, religious, economic domains, and eventually including education, science, and arts; and stratification systems composed of different strata or classes of individuals categorized by their prestige and perceived worth. At the same time, the structural base of earlier large societies—provided by linking nuclear families into lineages, clans and subclans, moieties, and other features of unilineal descent systems—was making its gradual retreat back to the more isolated nuclear family of modern societies (and, of course, the first human societies of hunter-gatherers).

With industrialization followed by postindustrialization, societies became large and complex, driven by markets and integrated by diverse institutional systems and cultural values, ideologies, laws, and norms. Humans had been able to endure the highly stratified societies that emerged after hunting and gathering through agrarianism because they had the cognitive capacities to understand the necessarily restrictive nature of structures limiting their option. As a result, individuals sought fulfillment in meeting needs in local kin units and the community. Still, the societies between hunting-gathering and postindustrialism were stages of societal evolution that were *not* well suited to humans' evolved nature because of the crushing constraints of stratification, the use of authority and coercion, and the limited individual freedoms; they were inherently unstable in the longer run, even as they tightly controlled their larger, stratified populations (Lenski 1964; Turner 1984, 1997, 2002). Urban centers and markets in feudal agrarianism and then with early industrialism and the increasing expansion of labor markets offered some opportunities for mobility and individualism. Urban areas in early industrialism were certainly grim places, and yet, they grew as human great apes sought some degree of freedom from feudal control. Still, humans were able to "prosper" in early industrial societies compared to the earlier societal formations (i.e., horticultural, herding, fishing, pastoral and then agrarian societies) that evolved after hunting-and-gathering formations because they had more choices even in the oppressive factory systems so well described by Engels (1845) in his portrayal of industrialism of the nineteenth century, before postindustrial societies fundamentally changed the scale and dynamism of human social organization.

The cognitive mapping abilities of humans increased during societal evolution to the point where it was possible to conceive of the boundaries of geographically large societies and even systems of societies organizing distinctive populations and revealing a wide diversity of cultural traits. Yet, for most of human social organization, even in late feudalism and early industrialism, the social universe of humans remained highly local, even as wars were fought at some distance with other populations, and even as industrial workers began to mobilize for conflict with industrialists. Workers could begin to see the whole world as a system, although for most individuals their focus was on the larger societal system that other animals on earth could not imagine. The evolution of ever more complex societies was only possible because individuals could see beyond community and see themselves and others in relation to other types of sociocultural formations, such as stratification systems (by class, ethnicity, and other categoric distinctions), pervasive institutional systems (economy, polity, religion, kinship, law, education, science, etc.), societal systems, and increasingly intersocietal systems. Humans' sense of self was oriented to identities lodged in categoric units, in corporate units making up communities and organizations, and in key roles in the divisions of labor of corporate units lodged inside of society-wide institutional systems (e.g., economy, education, polity, religion).

For the first time in human history, identities could begin to include societal-level identities as nationalism mobilized emotions about self in relation to "homelands" and their cultures. Mobility was now possible across the larger expanse of any given society or between societies. Interaction with strangers, non-kin, non-community, and even non-societal members was relatively easy for this evolved great ape with enhanced interpersonal skills and a large neocortex.

The features of this expanded scale and the scope of the community complex were no longer focused just on the local community. A capacity to cognitively perceive layers upon layers of differentiated social structures emerged. Nuclear families and local residential community remained a central anchor and focus of attention, but increasingly, humans could not only visualize but develop diverse emotional attachments to differentiated groups, communities, and organizations within institutional systems making up a society. They could see themselves and develop identities associated with not only categories in the stratification systems but virtually any of the many differentiated groups, organizations, communities, and societies of the new social universe. The elaboration of the interaction complex had allowed for incumbency in a diverse myriad of sociocultural formations, and the elaborated emotions complex allowed individuals to experience positive emotions in interaction rituals within many diverse social settings. In so doing, individuals could also meet at least some basic human need-states now driving the evolved psychological complex. Humans could move about within and across this myriad of social structures in ways never possible before. To a surprising degree, differentiation of social structures at any given level of social organization and further differentiation across levels of social structure created "opportunity spaces" to move and roam in ways individuals in horticultural and agrarian societies could not.

Even though elephants, members of the dolphin family, whales, and perhaps some species of highly intelligent birds can be mobile and interact in complex fission-fusion arrangements, no other animal can conceptualize the planet and all the sociocultural formations that organize almost 10 billion people, nor can they move about populations of strangers both inside their own communities and virtually every other social structure organizing their lives in a society or the lives of others in different societies. Humans are able to do this because of their great ape ancestry, allowing for weak ties, individualism, mobility, and cognitions focused on social structures beyond local groupings or kin units. Without this ape core, human nature would push for a highly parochial view of the universe and, hence, a reluctance to move beyond local groupings and territories. *Homo erectus* began this movement and expansion of hominins; humans have simply taken it further, creating societies that, as I will examine in the concluding chapter, are more compatible with human nature than any other societal form since hunting and gathering. Ironically, these societies are in danger of destroying much life on earth, including human life.

Conclusion: The Elaborated Community Complex

The critical difference between most mammals and the ancestors of great apes and hominins is that hominins are not naturally programmed to reckon the local group and kin units in the same way as most other mammals, including such closely related mammals as monkeys. The orientation to community was a critical preadaptation for macro societies for humans' eventual capacity to conceptualize such societies, feel comfortable in them, interact in them, meet needs in them, and derive positive emotions from them.

In contrast to most mammals, forming more stable grouping among hominins was the big evolutionary roadblock to their survival. Once it was surpassed, the elaboration machine began to kick in and grow subcortical and then neocortical areas in the human brain. With this cognitive and emotional growth came the capacity to extend the perspective of late hominins and early humans even further out, beyond what had been a large home range, if ever-larger societal-level formations populated by many millions of inhabitants organized in a complex and layered systems of groups, organizations, communities, institutional domains, and stratification were necessary for fitness. The cognitive capacities of humans had sufficiently expanded, as had their emotional and interpersonal capacities; and while these were not needed for reckoning the small world of hunter-gatherers, they represented a preadaptation for the more macro societies that developed over the past 12,000 years. The early elaborations of the brain among late hominins and early humans created the potential for humans to find solidarities and positive emotions in interactions within any group in any type of social structure, thereby allowing for at least minimal satisfaction of need-states in the elaborated psychology complex.

These elaborations converted original mapping of community as the most stable social structure organizing hominin groupings into what is really no longer a community complex but a "societal complex"—an ability to see the larger social structure, even from the vantage point of a local grouping. Humans have the capacity to "look up" from the group to organization to community to institutional domain to society and now even to intersocietal system and "see" at least the outlines of the larger sociocultural formation organizing their lives and activities, just as chimpanzees can see the community and its demography within which their lives are conducted. No other mammals can do this seemingly simple but actually quite complex act. It is only possible with the original wiring to reckon beyond the group, coupled to a brain that is three times as large as the brain of the LCAs of early hominins. Because emotional enhancement preceded cognitive growth in the neocortex, the ability to look beyond the group to ever more layers of social structure and their cultures was always accompanied by emotions that allowed humans to develop attachments to meso and macro social structures and their cultures and, if need be, to mobilize negative emotions to those structures constraining human nature. Thus, large-scale social structures constrain and integrate large populations, but like everything else in the social world, they are vulnerable to the arousal of negative emotions and collective action against constraints that go against humans' evolved nature.

Not only can humans see the expanded sociocultural space in which they live out their lives, they can also see themselves as an object in this space. Humans have the capacity and indeed the need to develop identities attached to particular levels of social structure and culture: corporate units, categoric units, and, at times, such macro corporate units as a whole society or system of societies (e.g., "I am an American"; "I am a European"), or a category across societies ("I am an Arab"; "I am a Jew"). Indeed, it is the ability to derive positive emotions from others' verifications of these senses of identity that allows humans to meet their most fundamental needs in different levels of social structures—at the level of roles, status, groups, organizations, communities, institutional systems, social categories, and whole societies. Again, at the same time, when such identities are not easily verified, individuals experience negative emotions. In these emotions lies the potential for tearing down structures that will not allow humans to meet the most powerful evolved need in their fundamental nature: the need to verify self.

The expanded interpersonal complex also gives humans not only the ability to identity-take and -make but, also, to role-take and -make, status-take and -make, structure-take and -make, situation-take and -make, culture-take and -make, and emotion-take and -make at any level of social structure, thereby enabling humans to "plug into" any social context and, potentially, meet all or most fundamental need-states. Moreover, as an inheritance from the LCAs of great apes and hominins, a "weak-tie" orientation also allows humans to move among strangers, something that is inevitable when living in a large and differentiated society. At the same time, humans' emotions complex and interaction complex give individuals the necessary tools to make friends and develop solidarities, sometimes short-lived but nonetheless emotionally gratifying, in order to function in almost any level of social structure and still meet the fundamental needs of the psychology complex. With these evolved complexes, humans can do what all great apes do: move about as individuals inside a more inclusive social structure—for chimpanzees, the community or home range and, for humans, the whole world of social structures organizing virtually all people on earth. Indeed, even without being able to speak the language of another population, the "language of emotions" based, in part, on universal sequences of emotional expressions is often enough to enable individuals to meet needs in distant lands. We are programmed by our community complex to understand the parameters and boundaries of our sociocultural environment, and we are enabled by our cognitive, emotions, and interpersonal complexes, as well as by our elaborated community complex, to function within the ever-shifting parameters of differentiated sociocultural systems that now organize humans.

In conclusion, then, Figure 11.1 summarizes the elaborated and evolved *community complex*. The inputs from our inherited nature are listed on the left side of the elaboration machine, and as they enter this elaboration machine, they are transformed and, indeed, elaborated into a *societal complex* or capacity to see both the immediate and larger sociocultural world in which we find ourselves. Only animals like humans can do so; and it is ever more remarkable for a large animal like humans to be able to organize mega societies without the instinctive programming evident among social insects. Whether it is beneficial to humans in the long run and, more importantly, to other life forms on earth is an open question—as the next, concluding chapter explores.

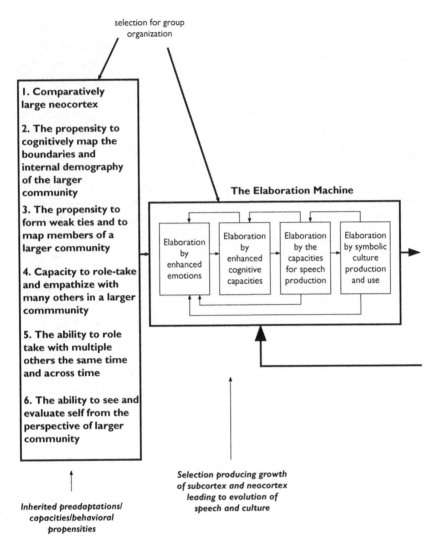

Figure 11.1 The Evolved Community Complex and Human Nature

1. Propensity of individuals to reckon and orient themselves to multiple levels of sociocultural formations, ranging from episodes of interaction in temporary encounters through groups within organizations and communities and, potentially, to larger institutional domains, stratification systems, societies, and intersocietal systems.

2. Capacity and propensity of individuals to see self from the structural locations and relevant culture to which they orient themselves, with identities most likely to be attached to (a) particular roles in corporate units and potentially the institutional domains in which they are embedded, (b) memberships in categoric units, particularly gender, ethnicity, and class but others as well, and (c) inhabitants of communities as well as memberships in particular groups and organizations within these communities.

3. Emotions generated by verifications of identities attached to various levels of social and cultural organization generate positive emotions that are rewarding, but also generate attachments and commitments to these various levels of organization as long as identities continue to be verified.

4. Capacity and propensity to role-take and -make and status-take and -make in efforts to verify self but also to understand expectations from locations in particular social structures for self and others, which, in turn, lead to structure-taking and -making, culture-taking and -making, and emotion-taking and -making. All these interpersonal practices help specify the levels of sociocultural organization that are most relevant to meeting psychological need-states of individuals, with meeting need-states increasing emotional attachments to various levels of social organization.

5. Ability and, at times, the preference to use the capacities outlined in point 4 to be among strangers and to engage them in necessary interactions at various levels of sociocultural organization.

6. The capacity and, at times, the preference made possible by points 4 and 5 to be mobile across several levels of social organization, particularly diverse types of corporate units (groups, organizations, and communities) in various institutional domains (e.g., kinship, economy, education, religion).

12

Human Nature and the Evolution of Mega Societies

Implications for Species and Personal Survival on Planet Earth

"I, a stranger and afraid
in a world I did not make."

A. E. Housman

Within the next two decades, the population on earth will exceed 10 billion inhabitants. If we go back to the very beginnings of humans, some 350,000 to 450,000 years ago, the population of all humanity, including species closely related to *Homo sapiens* (e.g., *Neanderthals, Denisovans,* and even *Homo naledi*), was small, not much more than a few million at best. Before leaving Africa, *Homo sapiens* were close to extinct; all over the earth, they suffered periodic declines as local environments changed (Amos and Hoffman 2010; Manica et al. 2007). How, then, did a vulnerable line of late hominins leading to *Homo sapiens,* who could barely get sufficiently organized to survive in the first place, ever create complex societies of billions of evolved great apes? If we can imagine ants and termites to be the size of humans, we can then imagine what such populous societies would do to the world's ecosystem as they sought sufficient resources to survive. Humans are not like ants; they are dramatically bigger and are organized by more complex societies than the social insects, while possessing brains that can create demands for resources and technologies to strip resources from the planet on an enormous scale and, then, dump the waste products of this consumption into the world's ecosystem. Humans are thus a threat to themselves and perhaps to all life forms on earth—with the exception, perhaps ironically, of the social insects that might indeed inherit the earth because of their numbers, small size, and efficient, genetically driven patterns of social organization.

The story of human evolution is one of a number of near extinctions; and then, over the past 10,000 years, population growth organized into societies that are best described by such terms as *mega, macro,* or *ultra societies.* Aside from insects, few life forms are organized by divisions of labor that can coordinate large populations; and humans are, by far, the biggest animals ever to do so, with the obvious consequences of environmental degradation—ranging from global warming and its consequences, through species depletion from effluents and pollutants, to simplification of the ecosystems' resources needed to sustain the diversity of life on earth. If humans had evolved from monkeys, whose societies are composed of local groups of related females and male (and sometimes female) hierarchies of dominance, the world would be a much safer place, although monkeys have been able to adapt to human urban centers in many parts of the less developed world often better than the humans who built these cities. Much like coyotes, monkeys, with their genetically governed group-level organization (versus pack organization for coyotes), are able to adapt to diverse environments, particularly to those with easily accessed sources of food. In contrast, humans carry a great ape "nature" dramatically magnified and elaborated by brains revealing a large palette of emotions, enhanced cognitive capacities, spoken language, and symbol-based culture. The five complexes outlined in the previous chapters allowed humans to organize ever more complex societies and, thereby, adapt to virtually any habitat. As this trend continued over the past 10,000 years, inhabitants of human societies are set to overpopulate the earth with potentially disastrous consequences.

Human Nature and the "Social Cages" Created during Societal Evolution

If we examine the inherited preadaptations and behavioral capacities and propensities of great apes and, by the logic of cladistic analysis, as a good proxy for the capacities and propensities of humans' hominin ancestors, then it is clear that weak social ties, lack of permanent groupings, orientation to community, and the interpersonal practices that sustained these features of great ape societies were distinctive features of humans' hominin ancestors. These same features also made hominins and then humans initially vulnerable to predation and other disasters once they left the protection of the forests. Yet, the comparatively large neocortex, the set of preadaptations and behavioral capacities and propensities inherited from our last common ancestors (LCAs) with great apes, coupled with primary emotions generated in the subcortex, comparatively large neocortex, and prewiring for language, can all be viewed as *preadaptations for macro societies*, if subject to further selection. The key was to survive open-country habitats and to let selection work on the inherited traits from the LCAs of great apes and hominins, remaking a vulnerable animal into the most dangerous animal on earth—not only to each other but also to most forms of life on earth. Human mega societies are the *Tyrannosaurus rex* of earth today, except human societies are much bigger and dramatically more ferocious in their effects on the environment and other life forms.

The evolved nature of humans—as outlined by the cognitive, emotions, psychological, interaction, and community complexes—enabled humans to create the first human society revolving around hunting and gathering bands composed of nuclear families. Despite being vulnerable to environmental changes, such bands were sufficiently flexible and even resilient in the face of environmental challenges to populate, sparsely, most of the earth and to reproduce. The hunting and gathering band was very much in tune with human nature because it allowed for individuals to be mobile, relatively unconstrained by kinship and band, social but not engulfed, group-living in only two basic groupings (nuclear family and band), and oriented to a home range of many square miles. Just enough structure to coordinate kin-based reproduction, gathering of plant life, and hunting as a source of the protein needed by their large brains was available. This basic societal formation survived for several hundred thousand years, albeit punctuated by episodic declines under rapid environmental changes (Manica et al. 2007; Amos and Hoffman 2010).

Humans had probably always settled down near abundant resources, such as oceans, lakes, and rivers with their plentiful fish as easy sources of protein. These semisettled populations began to grow, and perhaps growth would force splitting up of the larger group in search of new resources. Yet, this basic adaptation revolving around the hunting and gathering band of nuclear families was sufficient for *Homo sapiens* to survive and, in many ways, to be in tune with their great ape ancestry of loose social ties, individualism, and mobility, with just enough constraint by nuclear kinship and band to allow them to adapt to more open-country habitats and, later, more diverse habitats in most of the world. At some point, more populations settled down into more permanent communities and grew, forcing the creation of the more constraining social structures needed to coordinate the activities of the larger population.

Humans thus began to create their own sociocultural "cages" composed of more restrictive social structures and their cultures (Turner and Maryanski 1992, 2008). *Big Man societies* evolved near sources of water where a central leader and his allies often "owned" economic production (from fishing, domestication of animals, and plants) but were required to redistribute these resources to members of the population, thereby bringing prestige to the Big Man but also the expectations for and responsibility for redistribution of resources. Still, Big Man societies began to construct the "cage of power" that would be present in all subsequent societies. Pastoral societies of herding and plant cultivation did much the same, and then horticultural societies created the "cage of unilineal kinship" added onto the cage of power to coordinate larger numbers of settled gardeners who might not just hunt but also breed and domesticate livestock. Horticultural societies—revolving around gardening without the plow—could become quite large and reveal high concentrations of power and patterns of constant warfare with neighbors. So with advanced horticulture, societal evolution moved way beyond simple horticulture to societies revealing additional cages such as bureaucratized religious structures, state, military, stratification, and economic structures outside of kinship. With each level of structural elaboration and differentiation, ever larger populations could be sustained and controlled (Spencer 1874–1896). With full-blown agriculture, manorial estates (a new form of community organization revealing bureaucratic elements and inequalities of power and wealth) within various forms of feudalism were added to the structural mix inherited from advanced horticultural societies. Moreover, the increased productivity allowed for the differentiation of new institutional systems (e.g., education, law, art, science) beyond the initial core of kinship, religion, economy, polity, and law.

Ironically and perhaps inevitably, the inherited great ape nature as modified by the elaboration machine into the various complexes examined in previous chapters was, as a result, increasingly caged not only within new institutional systems but also within crushing stratification systems as well as varieties of community formations such as manorial estates, villages, towns, and large cities devoted to economic, political, and religious activities. Intersocietal and intercity state warfare were chronic, thus constantly re-creating the cage of power that engulfed individuals and their families.

If we review the various complexes constituting humans' evolved nature, it is clear that this increasing "caging" of evolved great apes went *against* the inherited nature of humans' hominin ancestors. Thus, these "evolved" social structures were not in tune with human nature, and hence, certainly not driven by this nature. Yet, it is likely that the evolved human nature outlined in the complexes exerts a constant counterpressure on social structures that restrict and cage what are still individualistic great apes, now charged up with more emotions than other animals and now capable of planning, thinking, and mobilizing to resist the restrictions imposed by social cages. Thus, as the cages of power, kinship, bureaucratized religion, city, manorial estates, conquest, economic exploitation, and stratification all evolved and allowed for ever larger societies, the nature of societies was not well aligned with humans' evolved nature, as were the first small societies composed of nuclear families in hunting and gathering bands. Hence, conflicts, revolts, conquests, migrations, and other patterns of human resistance all increased in the evolution of societies after hunting and gathering. As evolved apes, humans will seek to escape from cages, and they will gravitate toward sociocultural arrangements that allow them to meet the need-states outlined in the *evolved psychology complex*, to experience positive affect of the *evolved emotions complex*, to form flexible and loose social ties made possible by the *evolved interaction complex*, to achieve some degree of consistency and congruence among cognitions and emotions in the *evolved cognitive complex*, and to be more free and mobile within the social structures of the expanded *community complex*. Humans' evolved nature as expressed in the five complexes outlined in this book is most compatible with certain types of social structures, as we might recall from Chapter 1 (page 23) in Christakis's (2019: 13) list of traits that, in a wide variety of contexts, appears to be the "blueprint" for successful social relations. They include:

1. The capacity to have and recognize individual identity
2. Love for partners and offspring
3. Friendship
4. Cooperation
5. Preference for one's own group ("in-group bias")
6. Mild hierarchy (relative egalitarianism)
7. Social learning and teaching

What is particularly interesting in Christakis's list is that relatively slight variations from these organizational features disrupt social relations and make the social structures less viable, often being perceived by individuals as cages of constraint. Thus, while humans' evolved nature is quite complex and multidimensional, it evolved in the run-up to the nuclear family and hunter-gatherer band; the "blueprint" for social structures compatible with this nature is surprisingly precise and narrow. True, humans' evolved nature allows them to survive and reproduce in social structures that violate much of this nature—as is evident in the societies that have successively evolved since hunting and gathering. Yet, there will always be tension between humans and social structures deviating very far from Christakis's blueprint.

The reason that humans have been attracted to industrial and post-industrial societies is that the features of such societies—organization by markets, democratically elected power, religious freedom, nuclear kinship, open education, diverse systems of communities from rural villages to large urban centers, alternative economic opportunities, access to travel, recreational activities of one's choosing, and in general, more freedom, choice, mobility, and individualism—are more likely to build at least some structures using Christakis's "blueprint." If an evolved ape cannot be part of a hunting and gathering band, then postindustrial capitalism—for all its well-documented contradictions and flaws—is more in tune with human nature than Big Man, pastoral, horticultural, agrarian, and early industrial (with their well-documented exploitation) societies. Postindustrial societies are not utopias by any means, but they are better than all the alternatives since hunting and gathering. Without killing off most of the human population, going back to hunting and gathering is not an option. Indeed, a world of 10 billion hunter-gatherers would, given the resulting population densities, end up with humans hunting each other down.

Humans will be caged in complex societies unless environmental catastrophe suddenly kills off most humans—which is, of course, a real possibility. Yet, if cages in complex societies are sufficiently gilded and manage to use the "blueprint" outlined by Christakis, they are to be preferred to all likely alternatives. The pathologies of modern societies—environmental degeneration, class stratification, ethnic conflicts, worker exploitation, warfare, civil unrest, dictatorial governments, religious intolerance, family dysfunction, poverty and starvation, mental illness, and so on—are not directly driven by the complexes of human nature. Rather, these pathologies are the outcome of evolved forms of social organization responding to population growth more than pressures from humans' evolved nature. Thus, it is the evolved social structures and cultures of these contemporary societies, driven by sociocultural selection pressures rather than by human biology, that are now the problem for humans—an old theme in early sociology cast in a new light.

Mega societies, while providing some "space" and alternatives that are compatible with human nature, still restrict this nature to a high degree. Large, complex societies are always filled with contradictions, inequalities, injustices, abuses of power, discrimination, oppressive divisions of labor, and other restrictions necessary for such societies to be able to operate. Without population decline of an enormous magnitude, we are stuck with such societies. True, humans' biological nature has enabled the building of such societies, when it was necessary to organize larger populations, but ironically humans have built societies that can often thwart the realization of their fundamental nature, as outlined in the five complexes. The pathology inheres in what humans have had to create in order to survive in larger numbers, not so much in our nature as biological beings. While conflict, even warfare, and other pathologies certainly existed among hunter-gatherers, they were rather rare given the low population densities, the lack of inequality, the absence of consolidated of power, and the comparatively few restrictions on individuals of hunting and gathering bands. Hunting and gathering was the human Garden of Eden, but once populations began to settle down, grow, and become more complexly organized, humans had eaten the forbidden fruit that cast them out of their Garden of Eden, or more accurately, humans would destroy the Garden of Eden in order to build larger, more complex societies. Yet, it should be recognized that there is now an accumulating body of data revealing that even hunting and gathering societies often evidenced efforts of some to consolidate power and impose hierarchy on band members or even other bands. These efforts, it appears, were typically met by counter aggression, often involving the murder of those who sought to confine others in the cage of power. It is possible that this dialectic was quite common. As a result, norms of equality evolved and cooperative alliances formed in order to restrict efforts of any individual or subgroup to impose the cage of power (Boehm n.d.). Even in hunting and gathering, then, and certainly later in larger sociocultural formations, inequalities would arise and often led to resistance and conflict, typically motivating the formation of alliances against abusers of power and to sustain more equal social relations. In most cases, it appears that humans found a way to stabilize early societies and create cooperative structures that allowed humans to realize their evolved nature.

Why Do Humans Prefer Modernity?

At their genetic core, humans are just another a great ape. Great apes generally possess many weak over strong ties, mobility around communities, periodic gatherings where positive emotions are aroused to celebrate the community, sexual freedom and promiscuity, individualism, and options for choice. Indeed, the interaction complex among great apes and humans is set up to create and sustain social relations in larger communities, where many ties are weak but still capable of generating positive emotions and verifying identities and other human need-states. Humans are highly emotional animals—indeed, the most emotional animal on earth—and much of their interactions revolve around securing positive emotions, as individuals use the elements of the *interaction complex* to role-take and -make, status-take and -make, identity-take and -make, culture-take and -make, emotion-take and -make, structure-take and -make, and situation-take and -make in order to plug themselves into each other and the relevant social structures and cultures guiding their conduct. True, there is *constraint* inhering in these efforts, but it is still under quite a bit of voluntary control, while revealing flexibility as the situation requires. As long as people can realize a profit in exchanges of resources, especially the profit inhering in positive emotional arousal, humans will at least be content, even in complex social structures. Thus, added to the emotions and interaction complexes are the need-states of the *psychological complex* for verifying multiple levels of self, for just and fair outcomes in exchanges, for cognitive and emotional congruence, for a sense of efficacy, for a sense of group inclusion, for trust, and in general, for positive emotional arousal in interactions with others. If individuals can successfully complete interaction rituals and even totemize or symbolize their relations, they will experience positive emotions from group affiliations (Collins 1975, 2004). The cognitive complex provides the means to engage in what are often nuanced and multifaceted interactions that are regulated by social structures and culture but at the same time are sustained by individuals' abilities to organize and retrieve stocks of knowledge necessary to interact, verify self and other need-states, and experience positive emotions with respect to self, others, and situation.

The complexity of these processes in modern societies is actually quite staggering, but positive emotions can be experienced through the ability to navigate this complexity. Indeed, the evolved nature of humans is set up to deal with complexity and to interact with many different kinds of individuals in diverse types of situations. Humans are set up to experience positive emotions about others, self, situations, and sociocultural structures in which successful interactions occur. An individualistic, evolved great ape still remains individualistic, to some degree, and enjoys walking through what are now much more complex communities of various types of social structures while verifying self and experiencing positive emotions in interactions that any evolved chimpanzee would find gratifying. The greater freedoms of hunting and gathering are lost, but in their place arise many diverse opportunities in postindustrial societies, and therefore so many more opportunities to verify self and to meet other psychological needs, to use the large neocortex and what it stores, to experience a wide range of emotions made possible by the subcortex of human brains, and to employ the expanded skills at interpersonal alignment of human great apes, all revolving around efforts at arousing positive emotions.

Other properties of culture and social structure work against these opportunities—consolidation of power and authority in most social structures, restrictive norms, beliefs and ideologies, inequality in the distribution of resources and sometimes crushing stratification of members of a population into systems of privilege and poverty, economic exploitation, unequal access to key institutional systems such as education, and many other constraints. None of these restrictive conditions in postindustrial societies is as great when compared to conditions in the societies after hunting/gathering. What is now clear, is that human nature has evolved with a capacity to organize humans in societies vastly more complex than hunting and gathering. Natural selection perhaps overshot the mark on what was necessary to survive, but once the "elaboration machine" was created, perhaps such an outcome was inevitable. Complexity always generates niches, spaces, opportunities, and the like, especially in market systems with many diverse kinds of individuals and corporate units involved in organizing institutional activities. Humans' evolved nature is set up to take advantage of these openings in sociocultural space, thereby allowing humans to live and prosper in mega societies.

For years, I used to ask students in my classes when discussing hunting and gathering societies if they would like to go back and live in such a society. A large portion of the class almost always raised their hands to affirm that they would, but then when I started listing what they would *not* have, I could see students mentally taking their hands down. Camping is fun, for a time for most people, but as a lifestyle? Perhaps not, given the glitter of what can be done in postindustrial societies. Whenever I probed further, what became evident is that students, like most people in affluent postindustrial societies, like the varieties of options and experiences that are possible: the diverse opportunities to be mobile in space and across sociocultural formations, the diversity of persons to meet, the opportunities for self-enhancement, and so on. There are still cages, even occasional gulags, but many of them are gilded and allow individuals to meet basic need-states that are part of human nature and to utilize the other complexes that are part of human nature—all of which lead individuals to experience positive emotions, at least much of the time.

Human Nature and Species Survival

Thomas Malthus was the bearer of bad news in the eighteenth century, when he borrowed the imagery from the Bible on the "four horsemen of the apocalypse." Overpopulation eventually leads one or more of Malthus's "four horsemen" to ride through societies. The imagery of each rider mounted on a different-colored horse allowed Malthus to stress that war, pestilence, famine, disease, and death (which actually ends up being five horses) will strike societies where population exceeds the capacity to support itself. We now know, of course, that many conditions send one or more of the horsemen to ride through societies, causing death and destruction. Malthus's argument is still as relevant today as it was two centuries ago. It seems inevitable that humans will experience these four horsemen that bring along the fifth (death) in the coming decades, and on a global scale. Whether as a world nuclear or conventional war, as widespread terrorism using these weapons of mass destruction, as deadly germs traveling the globe at the speed of air travel, as widespread famine produced by global warming and environmental degradation, as rising seas and damage from weather that comes with global warming, or as any of the many other catastrophic events that now seem possible, human societies will be under siege. Indeed, when I started this book in 2019, the COVID-19 pandemic was not on the radar, but within six months, the coronavirus began to move across the globe. As has become evident, this virus has exposed the vulnerability of humans—not only their immune systems but also their institutional systems for dealing with the threat of a simple but clever virus to hide from the human immune system and, when discovered, trick this system into overreaction. The capacity to contain the virus rests as much on the institutional systems of a society—polity, economy, education, and law—as on a society's medical care systems or individuals' immune systems. The key to controlling pandemics is the ability of societies to control contact among members of the population, with some societies doing this well—for example, China (after an initial failure followed by a massively coercive response), South Korea, and Germany—and others doing very poorly. Indeed, the United States' response to the pandemic has been a lesson in poor preparedness and inadequate coordination of responses. Thus, high technology and advanced medical knowledge cannot protect humans from just one of the four horsemen, much less all four riding through and trampling the institutional systems of mega societies. The institutional grandeur and complexity of these societies is not as effective in stifling the horsemen as some might image.

Unfortunately, the institutional systems of societies are often the cause or at least a facilitator of the path of Malthus's horsemen through societies.

Malthus recognized that once humans began to experience "the good life" associated with societal development, they would reduce their rates of reproduction. This trend, evident in the modern world, has given humans a false sense of security. In many parts of the world, however, children are the only "social security" available to parents as they age—thus increasing the birth rates and size of the societies least able to protect their inhabitants from any of the four horsemen. The result is that birth rates are not coming down sufficiently in the affluent post-industrial world, generating much of the environmental degradation, while half or more of the world's population where the four horsemen are most likely to ride are still growing rapidly. In this global system of societies, the four horsemen can often ride through the gates of even affluent, postindustrial societies with greater ease than we might think. Thus, humans are on the verge of spoiling their gilded cages in the affluent part of the world, while those in the less-developed portions of the globe will suffer even more in the coming decades as the four horsemen ride again and again. War, terrorism, revolt, environmental degradation, rising waters, pandemics, and the like are still occurring every day somewhere on the globe and are likely to spread and become more global in the next decades.

Does human nature drive this impending conflagration? Evolved great apes, evidencing preferences for high individualism, choice and personal freedoms, and other traits outlined in the five complexes of human nature, can sometimes mobilize to deal with collective problems. At other times, however, humans resent the imposition of sociocultural cages necessary for dealing with any of the four horsemen. For example, COVID-19 virus pandemic exposed the United States as poorly organized at the federal level to respond to this kind of threat. Even as the vast majority of individuals isolated themselves and followed directives of state and local authorities, many political figures making key decisions along with many individual citizens asserting their "rights" to be individualistic have dramatically weakened the already chaotic response of government to the crisis posed by COVID-19. By the time this book is printed, we will know just how consequential the chaotic response of the United States and other societies, such as Mexico and Brazil, in the Western Hemisphere will be for human life and for damaging the economic systems on which human life depends.

The pandemic has given human mega societies a clear "shot across the bow" and warning that, if humans are unable to change the social structures and cultures of their societies because they either enjoy them or they do not have the power or knowledge to effect social change of such complex systems, then we are all likely to experience coming disruptions to social life. The basic problem inheres in the kinds of societies that are required to organize larger numbers of individuals because, contrary to what some argue rather glibly, alternatives to these kinds of societies that can organize billions of large and evolved great apes are not so easily implemented. We can all think of utopias, perhaps the ultimate being hunting and gathering, but we cannot go there unless the four horsemen kill off most humans, which, to be sure, is possible but not likely in the short run. For all the power of the cognitive complex to innovate, we are individualist at our biological core, loaded up with emotions and powerful psychological need-states that can work against collective responses to common threats. Still, the opposite has often been the case in the past, such as the mobilization of England and the United States to confront the Axis powers during World War II. For a time, of course, public opinion in World War II (and earlier in World War I) was resistant to military mobilization by the United States, once again reflecting the vulnerability to danger that humans' evolved nature can pose. In the case of the world wars, the United States eventually mobilized its population, military forces, and economy on an unprecedented level. Thus, we can ask, Will such be the case in the future? Our first real tests thus far in the twenty-first century, where viruses have spread across societal borders, do not inspire confidence. Moreover, the continued involvement of the United States in wars to quell terrorism has spent a great deal of money but does not appear to have led to any real resolution of the problems that started the wars in the first place. There are, of course, those who deny that the billions spent on war have been ineffective and that the responses to invasions by viruses have been adequately handled, and there are others who deny the potential consequences of climate change, ecological decay, and poverty-generating war. Such ideological differences in humans' views will make collective responses to threats difficult and less effective when undertaken.

Complex social structures and their cultures are difficult to change by directed action because of their scale and the entrenched interests that always resist large-scale and rapid change. When forced by events, societies can change, but they do not necessarily change in the direction required. And so, while elements of human nature can get in the way of change and collective responses, much more significant are *the inertial tendencies of social structures and culture*, especially in complex societal systems with large, diverse populations, that resist and work against change, even changes desperately needed by rapidly changing environmental conditions. Humans by their nature may be well adapted and even moderately content in the survival machines of the modern world—that is large, complex societies—but they are not able to effectively change these survival machines when threatened. Even as humans face the four horsemen down, it is often difficult to address the problems that will emerge in the next decades, even if it were possible to build a new survivor machine for 10 billion people. Thus, it may be that humans have created sociocultural monsters—that is, large, complex societies—that they can no longer control as we face potential extinction events like the first hominins on the human line as they adjusted to open-country habitats. George Stewart's book *Earth Abides*, published in 1947, is still in print because we humans perhaps recognize our future in its pages and, hence, the old book is still relevant to understanding the ecological and organizational problems in the future. It seems unlikely that humans would be forced to return to hunting and gathering, even with the magnitude of the changes now occurring in sociocultural and ecological domains. Ecological problems loom large, however, and will likely pose serious challenges to humans and their mega societies. Even with humans' elaborated *cognitive complex*, which has served the species so well over the past 400,000 years, humans may not be able to deal with the magnitude of the changes required over the next 200 years.

More Misery for Humans, Masquerading as Technological "Advancement"

Like much of modernity, great potential appears at hand in new communication technology to help humans realize "their inner great ape." Computers using the Internet, video game systems, cell phones, social networking tools, and many other gadgets and programs that allow texting, audio, and video interactions among humans would seem likely to enhance humans' evolved nature rather than work against this nature. Yet, ironically, much of this technology only gives us a surface happiness and in the end violates humans' evolved nature. What has emerged, surprisingly, is devices that do not always enhance interactions among real people, face to face. Instead, these new modes of communication often lead to a series of *faux* interactions among illusionary people/beasts inhabiting video games in an unreal fantasy world and faux interactions in faux communities of supposed "friends" who are prepared to "like" you and your posts until they choose not to like your faux presentations of self to a world whose boundaries are unknowable and filled with, it would seem, lots of people who are not "your friends" after all. If this were not enough, many humans have become more attached to their cell phones than to real people, and rely on their phones for the latest news, text, Facebook posts, and just about anything. In fact, humans now spend untold hours staring at the screen on their cell phone, even when trying to have an actual "conversation" with another co-present and very live human being.

Teenage suicide is at an all-time high in the United States and elsewhere in the modern world; old people feel alone without personal interaction with their grandchildren and children; career people seem otherwise engaged in cell phones and computers doing work that must be done, followed by compulsively checking emails and social media; and in general, there is relatively little actual face-to-face interaction in private and public places. I have watched couples in restaurants spend more time monitoring their respective cell phones than engaging in conversation; I have seen many people hardly talk to each over a table but instead pull up and show their partner a picture, and so it goes back and forth, a virtual interaction among two people facing each other but speaking virtually no words. I have been in conversations with others who look at their cell phones more than at me and who "must get this call or text" as if their lives depended on it, thus making the real other co-present feel degraded and rejected.

It may be that for all the mediated interaction, people today in postindustrial societies engage in less direct face-to-face interaction than was the case before the information revolution. When they do engage face to face, they split their time between their cell phone and the other in co-presence, thereby degrading the potential value of the interaction for themselves and for the person who is receiving only half their attention. There has, of course, been great discussion over such mediated interactions and what they mean for persons and societies. Viewed from the perspective of the complexes that constitute human nature, it seems to me that individuals are not getting sufficient emotional feedback from others; they are not meeting important need-states, such as affirmation of self, a sense of group inclusion in a real group, exchanges of valued resources, sense of efficacy, trust, and most importantly positive emotional exchanges. They are not deriving the benefits of being a member of a real community but only loneliness from interaction in a simulated community. Preoccupied with their cell phones and other media outlets, they still do not feel fulfilled, much like drug addicts who need their fix that, sadly, never seems to be enough.

The solution to individuals feeling this sense of being unfulfilled is surprisingly simple: engagement with more people, face to face. Humans, unlike most other mammals, are visually dominate and, thus, respond emotionally to communication in which individuals see each other as they speak. To increase face-to-face interaction would, however, involve putting down or aside those machines that offer the chimera of real human engagement but deliver only a facsimile of what humans really need. Simmel (1906) suggested that, while the modern, differentiated world was not as "warm" (or "cuddly") as more traditional societies, it offered other things that generate human well-being: choice, options, freedom, the ability to spend money in markets to meet a person's unique needs, and the capacity to join many different kinds of groups. I think that Simmel understood what weak social ties can do for an evolved ape: generate a sense of satisfaction and, when coupled with a few strong ties (close friends and family), modern market-driven societies can meet most human needs.

The irony is that marketing of information technologies has interrupted what should be natural propensities for humans: to say hello, engage in small talk, interact in dozens of encounters generating positive emotions each day, and with time and energy remaining to engage with personal talk in high-solidarity friendship and family groupings. Coupled with choice, options, freedoms, and other features of postindustrial societies, an evolved ape like humans can retain a sense of efficacy, individualism, and orientation to communities and, beyond, to macro-level social formations.

The COVID-19 pandemic has, however, revealed the potential of these somewhat alienating technologies to sustain personal ties. When efforts are made to stay in contact with close friends and family via Skype and Zoom, individuals can see and focus on each other. These interactions may not produce the same emotion-arousing rituals (Collins 1975, 2004) on seeing and hearing humans "in the flesh," but they are able to promote positive emotions among real people who can see each other face to face—albeit in a mediated video format. The potential for experiencing real human engagement is inherent in these media technologies, with improvements in the quality of the video and audio. Perhaps the ability to more easily shift focus on particular persons on new high-definition screens, something close to real personal interactions can occur. Even interactions that are less personal become "more real" and more quickly "personal" when individuals can see each other and their expressive gestures in almost any context. Just making inter prsonal contact with, for example, the staff a pick-up restaurant, workplace, library desk, or any other interaction with acquaintances but not intimates is still highly gratifying for an evolved species used to weak ties. These technologies are not bound by local community, but they will have meaning within some community or more inclusive social formation, because this is how humans are programmed by the community complex. Community can be dramatically expanded to any place in the world, which allows technologies now available but not normally used for weak *and* strong ties to be reinforced by face-to-face interactions that can meet the need-states of humans' psychology using the capacities of the emotions and interaction complexes. For example, in comparing the times that I talked with one of my granddaughters in a French university by phone and then later by Skype (transferred to my big OLED TV screen), visual engagement on a big, human-sized screen certainly enhanced the emotional feelings. Such is the power of the visual sense modality. Thus, if humans in the modern world wish to take advantage of the options and choices emphasized by Simmel, current media technologies can greatly expand them. The key to this expansion is to begin to limit use of the more addictive technologies—that is, game systems attached to the Internet, even when they reach across the globe, and cell phones where individuals are constantly monitoring (and often evaluating) themselves in "likes," texts, emojis, and numbers of "friends" in a network that is too large to offer much of what humans need. This accumulated distortion of human relations by devices begins in childhood, accelerates into a true addiction in the teen years (as do other addictions), and becomes routinized in the adult years. For all the convenience of these technologies (e.g., to phone and text anywhere, anytime), they have not made humans happier, unless persons exercise discipline over the intrusion of this technology into life. The complexes of human nature allow us to have this addiction and to function, but it is now becoming clear that these technologies do not promote positive emotions if they are used as a substitute for face-to-face interactions in meeting the needs of the psychology and emotions complexes. Indeed, humans today often do not exercise their sophisticated capacities in the interpersonal complex to engage in a wide range of weak and strong face-to-face interpersonal ties, and in so doing, they are throwing away much of their great ape legacy by failing to activate their interpersonal capacities to arouse positive emotions and meet fundamental psychological needs.

Recapturing Our Humanity in Complex Societies

Many of the big structural and cultural restrictions on humans in big societies will be more difficult to overcome, but there are *personal choices that individuals can make* that can recapture what any hunter-gatherer has, even in the context of constraint. The first task of societal structural reconstruction is to attack inequalities in the distribution of power, material resources, and symbolic resources such as honor, dignity, and prestige. This is the place to begin the reconstruction of societies, unless it is too late. Still, *at the level of the person*, the analysis of human nature in the pages of this book tells each of us what we must do to make our lives more gratifying: *increase rates of interaction with real people* in diverse contexts, mixed with both weak-tie and strong-tie interactions and group affiliations revealing varying levels of attachment and solidarity. Humans are, like all great apes, individualists, but we are not designed by natural selection to be either "loners" or "addicts" to communication devices. Humans are designed to use the communication skills and interpersonal capacities that natural selection worked so hard to install so that we can experience positive emotions and solidarities with real rather than mediated and often artificial others. Soldiers and warriors in video games, Alexa and Siri, friends on Facebook, exchanges of texts involving emojis, and all the other ways that humans mediate their interactions are *not* the kinds of interactions that humans need. These media interactions can be fun to play with, useful for keeping in touch, and essential in a busy life, but as I have perhaps overemphasized, they do not deliver what we humans need most: experiencing positive emotional arousal from meeting humans' psychological need-states that arise from using humans' elaborated cognitive, emotional, and interpersonal capacities to interact with real people in diverse social groupings within what is now vastly extended communities that make up the multiple dimensions and layers of modern human societies.

Because humans do not have strong bioprogrammers like most other mammals for kinship and for many strong ties, these kinds of kin and personal relations need to be actively constructed, and reconstructed, through the model on "interaction ritual chains" outlined by Collins (see Figure 10.2 on page 232). Similarly, weaker ties with strangers, colleagues, neighbors, and other forms of social relations also require interaction rituals to smooth the flow of interaction, but this process of interacting face to face arouses positive emotions and allows humans to exercise their interpersonal skills and, in so doing, meet psychological needs and affirm the sense of a person in a community. Christakis's "blueprint" discussed in Chapter 1 and listed again earlier in this chapter lays out the conditions where interactions with others are most likely to generate the social ties and social structures that bring human gratification. Deviation from this blueprint works against human nature. Therefore, once we know (1) the basics of humans' inherited nature, as elaborated by expanded emotions and cognitive capacities, speech production, and symbolic culture, and (2) the basic type of social structure that allows individuals to interact in ways that reinforce humans' evolved nature, we then have the more complete blueprint for what makes humans content and fulfilled and what makes large-scale societies more viable. If a preponderance of social units—from groups, through organizations and communities, to institutional domains and societies as a whole—in which interaction is ordered by Christakis's blueprint, then the complexes outlined in Chapters 7–11 or, if one prefers, the elements of *personhood* outlined by Christian Smith in Chapter 1 (pages 20 to 22), can be activated in ways that make humans happy and, in so doing, make small- and large-scale societies viable.

The COVID-19 pandemic has exposed the weakness of some societies, such as the United States, and the importance of social relations in social structures built from Christakis's blueprint in generating happiness. In my neighborhood, which is blocked off by a gate surrounding a golf course, I have never before seen in my years living and writing from my study (where I can see both the golf course and the street in front of my house) so many people walking alone with their thoughts but, I suspect, thoughts about their life and social relations (because I also see smiles on their faces in a difficult time). I also see something that I have never seen living here for over a decade: so many families out walking together with pets in tow while yakking, skipping, laughing, and just enjoying each other's company as a family. When I take my daily bike ride during this isolation phase of dealing with the pandemic, I experience a sense of community from all of the hellos, waves, and comments as I pass by walkers, other bikers, and people in golf carts. I know virtually none of these people, but these short rituals of mutual being and place are the very thing an evolved chimpanzee needs: a sense of being part of a community. So, the pandemic may have shown people what their future can be like, if some forms of mediated relations are downplayed (cell phones) and other forms of mediated relations with family and friends are enhanced through dramatically improved versions of Skype, Zoom, and other such platforms. Additionally, if these platforms are fed through larger, high-definition screens on TVs rather than small computer screens, the emotion-arousal effects will be that much greater (for, there is something emotionally uplifting in making human's face full sized on the big screen rather than seen as a miniaturized person on the computer screen). Coupled with the daily stories in the three newspapers that I read religiously every morning, it is difficult not to be impressed at how people in so many walks of life have used media to sustain or create new relations and to coordinate complex activities, such as concerts, plays, graduation ceremonies, and other forms of virtual gathering for creative and emotional purposes (and of course, for work, as well). We are using our cognitive complex to activate all the other complexes and to realize humans' fundamental nature. If these mediated relations can supplement real face-to-face relations in social units following the blueprint and activating interaction rituals, then modern societies can help humans be what they were evolved to be. Of course, inequalities, prejudices and discrimination, abuses of power, cultural conflicts, and other disintegrative forces that have always been with humans, even in simple societies but always in complex societies, can work against what is needed. Thus, much needs to be done at the political and macro-structural level to mitigate the tension-generating machine that inheres in all modern societies and reveals high levels of inequality and other disintegrative forces. For individuals at the more micro level of inter-personal behavior in smaller-scale social structures, there is an easier path to realizing the human potential for well-being: use the capacities inhering in the complexes of human nature to engage in interaction rituals with diverse others face to face, supplemented by new forms of video-driven face-to-face interactions. The key is to use what several million years of natural selection worked so hard to install in our genomes.

Bibliography

Abrutyn, Seth and Anna S. Mueller. 2016. "When Too Much Integration and he New Evolutionary Sociology: Recent and Revitalized Theoretical and Methodological Approaches Regulation Hurt: Re-Envisioning Durkheim's Altruistic Suicide." *Society and Mental Health* 6 (1): 56–71.

____. 2018. "Towards a Cultural-Structural Theory of Suicide: Examining Excessive Regulation and Its Discontents." *Sociological Theory* 36 (1): 48–66.

Alcock, John. 2001. *The Triumph of Sociobiology*. New York: Oxford University Press.

Amos, W. and J. I. Hoffman. 2010. "Evidence that Two Main Bottleneck Events Shaped Modern Human Genetic Diversity." *Proceedings of the Royal Society* 277: 131–137.

Andrews, P. 2019. "Last Common Ancestor of Apes and Humans: Morphology and Environment." *Folia Primatol* 91: 122–148.

Andrews, Peter and Lawrence Martin. 1987. "Cladistic Relationships of Exant and Fossil Hominoids." *Journal of Human Evolution* 16: 101–118.

Ardesch, J. Dirk, H. L. H. Scholtens, L. Li, T. M. Preuss, J. K. Rilling, and M.n P. van den Heuvel. 2019. "Evolutionary Expansion of Connectivity between Multimodal Association Areas in the Human Brain Compared with Chimpanzees." *Proceedings of the National Academy of Science* 116 (14): 7101–7106.

Ardrey, Robert. 1961. *African Genesis: A Personal Investigation into the Animal Origins and Nature of Nan*. New York: Delta Books.

____. 1966. *The Territorial Imperative*. New York: Atheneum.

____. 1970. *The Social Contract: A Personal Inquiry into the Evolutionary Sources of Order and Disorder*. New York: Atheneum.

Bachofen, J. J. 1861 [1931]. Das Mutterrecht. In *The Making of Man: An Outline of Anthropology*, edited by V. F. Calverton. New York: Modern Library, Calverton.

____. 1967. In *Myth, Religion, and Mother-Right: Selected Writings of J. J. Bachofen*, translated by R. Manheim. Princeton, NJ: Princeton University Press.

Baizer, J. S., J. F. Baker, K. Haas, and R. Lima. 2007. "Neurochemical Organization of the Nucleus *Paramedinaus dorsalis* in the Human Brain." *Brain Research* 1176, 45–52.

Baldwin, P. J. 1979. *The Natural History of the Chimpanzee (Pan troglodytes verus) at Mt. Assirik, Senegal*. Ph.D. Thesis, University of Stirling, Scotland.

Baldwin, P. W. C. McGrew and C. Tutin. 1982. "Wide-ranging Chimpanzees at Mt. Assirik, Senegal." *International Journal of Primatology* 3: 367–385.

Barger, Nicole, Kari L. Hanson, Kate Teffer, Natealie M. Schedker-Ahmed, and Katernia Semendeferi. 2014. "Evidence for Evolutionary Specialization in Human Limbic

Structures." *Frontiers in Human Neurosciences* 8 (article 277): 1–17. http://dx.doi. org/10.3389/fnhum.2014.00277.

Barger, Nicole, Lisa Stefanacci, Cynthia Schumann, Chet C. Sherwood, Jacopo Annese, John M. Allman, Joseph A. Buckwalter, Patrick R. Hof, and Katerina Semendeferi. 2012. "Neuronal Populations in the Basolateral Nuclei of the Amygdala Are Differentially Increased in Humans Compared With Apes: A Stereological Study." *Journal of Comparative Neurology* 520: 3035–3054.

Barger, Nicole, Lisa Stefanacci, and Katerina Semendeferi. 2007. "A Comparative Volumetric Analysis of the Amygdaloid Complex and Basolateral Division of the Human and Ape Brain." *American Journal of Physical Anthropology* 134: 392–403.

Barkow, Jerome, H. Leda Cosmides, and John Tooby, Eds. 1992. *The Adapted Mind: Evolutionary Psychology and the Generation of Culture*. New York: Oxford University Press.

Beckoff, Marc and Jessica Pierce. 2009. *Wild Justice: The Moral Lives of Animals*. Chicago, IL: University of Chicago Press.

Bellah, Robert. 2011. *Religion in Human Evolution: From the Palaeolithic to the Axial Age*. Cambridge, MA: Harvard University Press.

Berger, Joseph. 1958. *Relations Between Performance, Rewards, and Action Opportunities in Small Groups*. Unpublished Ph.D. dissertation, Harvard University.

———. 1972. "Directions in Expectation States Research." In *Status Generalization: New Theory and Research*, edited by M. Webster and M. Foschi. Stanford, CA: Stanford University Press.

———. 1988. "Directions in Expectation States Research." Pp. 450–474 in *Status Generalization: New Theory and Research*, edited by M. J. Webster and M. Foschi. Stanford, CA: Stanford University Press.

Berger, Joseph, M. H. Fisek, R. Z. Norman, and M. Zelditch, Jr. 1977. *Status Characteristics in Social Interaction*. New York: Elsevier.

Berger, Joseph and Morris Zelditch. 1985. *Status, Rewards, and Influence*. San Francisco: Jossey-Bass.

Bidney, David. 1944. "The Concept of Culture and Some Cultural Fallacies." *American Anthropologist* 46: 30–44.

———. 1947. "Human Nature and the Cultural Process." *American Anthropologist* 49: 375–399.

Blau, Peter M. 1964. *Exchange and Power in Social Life*. New York: Wiley.

Boehm, Christopher. 2013. *Moral Origins: The Evolution of Virtue, Altruism, and Shame*. New York: Basic Books.

———. n.d. *Ancestral 'Ambivalence,' Tribal Confederations, and The Evolution of Democracy*. Submitted for publication 2020.

Bornkessel-Schlesewsky, I., M. Schlesewsky, S. L. Small, and J. P. Rauschecker. 2014. "Neurobiological Roots of Language in Primate Audition: Common Computational Properties." *Trends in Cognitive Science* 19: 142–150.

Brosnan, Sarah F. 2006. "Nonhuman Species Reactions to Inequity and Their Implications for Fairness." *Social Justice Research* 19.

———. 2014. "Evolution of Responses to (Un)Fairness." *Science* 346 (6207): 125776.

Brosnan, Sarah F. and Frans B. M. de Waal. 2003. "Animal Behaviour: Fair Refusal by Capuchin Monkeys." *Nature* 428: 128–140.

Brosnan, Sarah F., Hillary C. Schiff, and Frans B. M. de Waal. 2005. "Tolerance for Inequity May Increase with Social Closeness in Chimpanzees." *Proceedings of the Royal Society of London* 272: 253–258.

Brothers, Leslie. 1997. *Friday's Footprint: How Society Shapes the Human Mind*. Oxford, UK: Oxford University Press.

Brown, Donald. 1991. *Human Universals*. New York: McGraw-Hill.

Burghardt, Gordon M. 2005. *The Genesis of Play: Testing the Limits*. Cambridge, MA: M.I.T. Press.

Burke, Peter J. 1989. "Sex Differences in Human Mate Preference: Evolutionary Hypotheses Testing in 37 Cultures." Behavioral and Brain *Science* 12: 1–49.

____. 2021. The Microsociology of Self and Identity. In *Theoretical Sociology: The Future of a Disciplinary Foundation*, edited by S. Abrutyn and K. McCaffree. New York and London: Routledge.

Burke, Peter J. and Jan E. Stets. 2009. *Identity Theory*. New York: Oxford University Press.

Buss, David and David P. Schmitt. 1993. "Sexual Strategies Theory: An Evolutionary Perspective on Human Mating." *Psychological Review* 100: 204–242.

Buss, David M. 2016. *Evolutionary Psychology: The Science of the Human Mind*: Fifth Edition. New York: Free Press.

Buss, David, M. Abbott, A. Angleitner, A. Asherian, A. Biaggio, and 45 authors. 1990. "International Preferences in Selecting Mates: A Study of 37 Cultures." *Journal of Cross-Cultural Psychology* 21: 5–47.

Call, Josep and Michael Tomasello. 2008. "Do Chimpanzees Have a Theory of Mind: 30 Years Later." *Trends in Cognitive Science* 12: 187–192.

Carey, A. D. and Joseph Lopreato. 1995. "The Evolutionary Demography of the Fertility-Mortality Quasi-Equilibrium." *Population and Development Review* 21: 726–737.

Chapais, Bernard. 2008. *Primeval Kinship: How Pair-Bonding Gave Birth to the Human Group*. Cambridge, MA: Harvard University Press.

Christakis, Nicholas A. 2019. *Blueprint: The Evolutionary Origin of a Good Society*. New York: Little, Brown, Spark.

Coleman, Daniel. 2005. *Emotional Intelligence*. New York: Bantam Books.

Coleman, James S. 1990. *Foundations of Social Theory*. Cambridge MA: Belknap.

Clark, Amy. 2012. "Embodied, Embedded, and Extended Cognition." Pp. 275–292 in *The Cambridge Handbook of Cognitive Science*, edited by K. Frankish and W. Ramsey. Cambridge: University of Cambridge Press.

Collins, Randall. 1975. *Conflict Sociology: Toward an Explanatory Science*. New York: Academic Press.

____. 1981. On the Micro-Foundations of Macro-Sociology. *American Journal of Sociology*, 86, 984–1014.

____. 2004. *Interaction Ritual Chains*. Princeton, NJ: Princeton University Press.

____. 2008. *Violence: A Micro-sociological Theory*. Princeton, NJ: Princeton University Press.

Cooley, Charles Horton. 1902 [1964]. *Human Nature and the Social Order*. New York: Schocken Books.

Corballis, Michael C. 2009. "The Gestural Origins of Language." *WIRE's Cognitive Science* 1 (January/February): 2–7.

____. 2010. "Did Language Evolve Before Speech?" Pp. 115–123 in *The Evolution of Human Language: Biolinguistic Perspectives*, edited by R. K. Larson, V. Deprez, and H. Yamakido. Cambridge: Cambridge University Press.

_____. 2017a. "Language Evolution: A Changing Perspective." *Trends in Cognitive Sciences* 21 (4): 229–236.

_____. 2017b. "A Word in the Hand: The Gestural Origins of Language." Pp. 199–218 in *Neural Mechanisms of Language*, edited by M. Mody. New York: Springer.

Cords, Marina. 2012. "The Behavior, Ecology and Social Evolution of Cercopithecine Monkeys." Pp. 91-122 in *The Evolution of Primate Societies*, edited by J. Call, P. Kappeler, R. Palmobit, J. Silk. Chicago IL: University of Chicago Press.

Cosmides, Leda and John Tooby. 1992. "Cognitive Adaptations for Social Exchange." Pp. 163–228 in *The Adapted Mind: Evolutionary Psychology and the Generation of Culture*, edited by J. H. Barkow, L. Cosmides, and J. Tooby. New York: Oxford University Press.

Crespi, Bernard, Silven Read, and Peter Hurd. 2017. "Segregating Polymorphisms of FOXP2 Are Associated with Measures of Inner Speech, Speech Fluency and Strength of Handedness in a Healthy Population." *Brain and Language* 173: 33–44.

Crockford, Catherine, Roman M. Witting, and Klaus Zuberbuhler. 2017. "*Vocalizing in Chimpanzees Is Influenced by Social-Cognitive Processes.*" *Science Advances* 3 (11): e1701742. https://doi.org/10.1126/sciadv.1701742.

Damasio, Antonio. 1994. *Descartes' Error: Emotion, Reason, and the Human Brain.* New York: G. P. Putman.

Damasio, Antonio and Norman Geschwind. 1984. "The Neural Basis of Language." *Annual Review of Neuroscience* 7: 127–147.

Damasio, Antonio R. 2000. *The Feeling of What Happens: Body, Emotion and the Making of Consciousness.* London: Heinemann.

Darwin, Charles. 1859 [1958]. *On The Origins of Species, By Means of Natural Selection.* New York: New American Library.

_____. 1871 [1875]. *The Descent of Man and Selection in Relation to Sex.* New York: D. Appleton and Co.

_____. 1872. *The Expression of the Emotions in Man and Animals.* London, UK: John Murray.

Dawkins, Richard. 1976. *The Selfish Gene.* Oxford, UK: Oxford University Press.

_____. 2006. *The God Delusion.* New York: Houghton Mifflin.

Deacon, Terrance D. 2009. "Relaxed Selection and the Role of Epigenesis in the Evolution of Language." Pp. 740–760 in *Oxford Handbook of Developmental Behavioral Neuroscience*, edited by Mark Blumberg, John Freeman, and Scott Robinson. New York: Oxford University Press.

Dean, Christopher and Meave Leakey. 2004. "Enamel and Dentine Development and the Life History Profile of *Victoriapithecus macinnesi* from Moboko Island, Kenya." *Annals of anatomy* 28: 405–415.

Dewey, John. 1922. *Human Nature and Conduct: An Introduction to Psychology.* New York: Henry Hold and Co.

Dirks, W. and J. E. Bowman. 2007. "Life History Theory and Dental Development in Dour Species of Catarrhine Primates." *Journal of Human Evolution* 53: 309–320.

Donald, Merlin. 1991. *Origins of the Modern Minds: Three Stages of Evolution of Culture and Cognition.* Cambridge, MA: Harvard University Press.

_____. 2001. *A Mind So Rare: The Evolution of Human Consciousness.* New York: Norton.

Durkheim, Émile. 1893 [1997]. *The Division of Labor in Society.* New York: Free Press.

_____. 1912 [1965]. *The Elementary Forms of the Religious Life.* New York: The Free Press.

Dunbar, Robin. 1984. *Grooming, Gossip and the Evolution of Language.* London: Faber and Faber.

Ebil-Eibesfeldt, Iranaus. 1996. *Love and Hate: The Natural History of Behavior Patterns.* New York: Aldine de Gruyter.

Eccles, John C. 1989. *Evolution of the Brain: Creation of Self.* London: Routledge.

Eisenstadt, S. N. 1964. "Social Change, Differentiation and Evolution." *American Sociological Review* 29: 375–386.

Ekman, Paul. 1984. "Expression and the Nature of Emotion." Pp. 319–343 in *Approaches to Emotion,* edited by K. Scherer and P. Ekman. Hillsdale, NJ: Lawrence Erlbaum.

Ember, Carol R. 1978. "Myths about Hunter-Gatherers." *Ethnology* 17: 439–448.

Emde, Robert N. 1962. "Level of Meaning for infant Emotions: A Biosocial View." Pp. 1–37 in *Development of Cognition, Affected and Social Relations,* edited by W. A. Collins. Hillsdale, NJ: Lawrence Erlbaum.

Emerson, Richard. 1962. "Power-Dependence Relations." *American Sociological Review* 17: 31–41.

Enard, Wolfgang. 2007. The Evolution of Broca's Area, IBRO History of Neuroscience. http://www.ibro.info/Pub/Pub_Main_Display.asp?LC_Docs_ID=3145.

_____. 2016. "Evolution of the Primate Brain." Pp. 1495–1525 in *Handbook of Paleoanthropology,* edited by W. Henke and I. Tattersall. New York: Springer.

Enard, Wolfgang, Philipp Khaitovich, Joachim Klose, Sebastian Zollner, Florian Heissig, Patrick Giavalisco, Kay Nieselt-Struwe, Elaine Muchmore, Ajit Varki, Rivka Ravid, Gaby M. Doxiadis, Ronald E. Bonttrop, and Svante Paabo. 2002. "Intra-and Interspecific Variation in Primate Gene Expression Patterns." *Science* 296: 340–342.

Enard, Wolfgang, Molly Przeworski, Simon E. Fisher, Cecilia S. L. Lai, Victor Wiebe, Takashi Kitano, Anthony P. Monaco, and Svante Paabo. 2002. "Molecular Evolution of FOXP2, A Gene Involved in Speech and Language." *Nature* 418: 869–872.

Engels, Friedrich. 1845. *The Conditions of the Working Classes in 1844* (in German). Leipzig: Otto Wigand.

Fahy, Geraldine, M. Richards, J. Riedel, J-J. Hubin, and C. Boesch. 2013. "Stable Isotope Evidence of Mean Eating and Hunting Specialization in Adult Male Chimpanzees." *Proceedings of the National Academy of Sciences* 110: 5829–5833.

Falk, Dean. 2000. *Primate Diversity.* New York: Norton.

_____. 2007. *The Evolution of Broca's Area,* IBRO History of Neuroscience. [www.ibro.info/Pub/Pub_Main_Display.asp?LC_Docs_ID=3145].

Ferrari, Pier Francesco and Giacomo Rizzolatti, Eds. 2015. *New Frontiers in Mirror Neuron Research.* Oxford, UK: Oxford University Press.

Festinger, Leon. 1957 [1962]. *A Theory of Cognitive Dissonance.* Evanston, IL: Row Peterson, reprinted by Stanford University Press in 1962.

Fisher, R. A. 1930. *The Genetical Theory of Natural Selection.* Oxford, UK: Clarendon.

Forey, P. L. et. al. 1992. *Cladistics: A Practical Course in Systematics.* Oxford, UK: Clarendon Press.

_____. 1994. Systematics Association Special Volume, No. 50, *Systematics and Conservation of Evaluations.* London: June 17–20, Oxford: Oxford University Press.

Franks, David D. 2010. *Neurosociology: The Nexus between Neuroscience and Social Psychology.* New York: Springer.

Franks, David D. and Jonathan H. Turner. 2012. *Handbook of Neurosociology.* New York: Springer.

Freud, Sigmund. 1899 [1900]. *The Interpretation of Dreams*. London: Hogarth Press.

Gallup, Gordon, Steven Platek, and Kirstina Spaulding. 2014. "The Nature of Recognition Revisited." *Trends in Cognitive Science* 18: 57–58.

Gallup, Gordon G., Jr. 1970. "Chimpanzees: Self-Recognition." *Science* 167: 86–87.

———. 1979. *Self-Recognition in Chimpanzees and Man: A Developmental and Comparative Perspective*. New York: Plenum Press.

———. 1982. "Self-Awareness and the Emergence of Mind in Primates." *American Journal of Primatology* 2: 237–248.

Garfinkel, Harold. 1967. *Studies in Ethnomethodology*. Englewood Cliffs, NJ: Prentice-Hall.

Garner, R. L. 1896. *Gorillas and Chimpanzees*. London: Osgood, McIlaine and Co.

Gazzaniga, Michael S. and Charlotte S. Smylie. 1990. "Hemisphere Mechanisms Controlling Voluntary and Spontaneous Mechanisms." *Annual Review of Neurology* 13: 536–540.

Geertz, Clifford. 1953. "Universal Categories of Culture." Pp. 507–523 in *Anthropology Today: An Encyclopedic Inventory*. Chicago: University of Chicago Press.

———. 1959. "Common Humanity and Diverse Cultures." Pp. 245–284 in *The Human Meaning of the Social Sciences*, edited by Daniel Lerner. New York: Meridian.

———. 1965. "The Impact of the Concept of Culture on the Concept of Man." Pp. 93–118 in *New Views of the Nature of Man*, edited by J. R. Piatt. Chicago: University of Chicago Press.

Gergely, Gyorgy and Gergely Csibra. 2006. "Sylvia's Recipe: The Role of Imitation and Pedagogy." Pp. 229–255 in *The Transmission of Cultural Knowledge*, edited by N. J. Enfield and S. C. Levinson. Oxford, UK: Berg Press.

Geschwind, Norman. 1965a. "Disconnection Syndromes in Animals and Man, Part I." *Brain* 88: 237–294.

———. 1965b. "Disconnection Syndromes in Animals and Man, Part II." *Brain* 88: 585–644.

———. 1985. "Implications for Evolution, Genetics, and Clinical Syndromes." Pp. 247–278 in *Cerebral Lateralization in Non-Human Species*, edited by S. D. Glick. Orlando: Academic.

Geschwind, Norman and Antonio Damasio. 1984. "The Neural Basis of Language." *Annual Review of Neuroscience* 7: 127–147.

Giddens, Anthony. 1984. *The Constitution of Society*. Berkeley CA: University of California Press.

Goffman, Erving. 1959. *The Presentation of Self in Everyday Life*. New York: Penguin.

———. 1967. *Interaction Ritual*. Garden City, NY: Anchor Books.

———. 1974. *Frame Analysis: An Essay on the Organization of Experience*. New York: Harper and Row.

Guttelmann, David, Josep Call, and Michael Tomasello. 2009. "Do Great Apes Use Emotional Expressions to Infer Desires?" *Developmental Science* 12 (5): 688–698.

Hamilton, William D. 1964. "The Genetical Evolution of Social Behaviour" *I and II*. *Journal of Theoretical Biology* 7 (1–16): 17–52.

Hare, Brian. 2006. "Chimpanzees Deceive a Human Competitor by Hiding." *Cognition* 101: 495–514.

———. 2011. "From Hominoid to Hominid Mind: What Changed and Why?" *Annual Review of Anthropology* 40: 293–309.

Hare, Brian, Josep Call, and Michael Tomasello. 2001. "Do Chimpanzees Know What Conspecifics Know?" *Animal Behavior* 61: 139–151.

Hechter, Michael. 1987. *Principles of Group Solidarity*. Berkeley: University of California Press.

Heider, Fritz. 1946. "Attitudes and Cognitive Dissonance." *Journal of Psychology* 21: 107–112.

____. 1958. *The Psychology of Interpersonal Relations*. New York: Wiley.

Hernandez-Aguilar, R. A., J. Moore , and T. R. Pikering. 2007. Savanna Chimpanzees Use Tools to Harvest the Underground Storage Organs of Plants. *Proceedings of the Natural Academy of Sciences*, 104, 19210–19213.

Hobbes, Thomas. 1651. *Leviathan*. Online version Pacific Publishing Studio, *amazon.com/books*.

Hochschild, Arlie R. 1979. "Emotion Work, Feeling Rules and Social Structure." *American Journal of Sociology* 85: 551–575.

Homans, George C. 1962. *Social Behavior: Its Elementary Forms*. New York: Harcourt, Brace, Jovanovich.

____. 1975. *Social Behavior: Its Elementary Forms*: Revised Edition. New York: Harcourt, Brace, and Jovanovich.

Holoway, Ralph L. 2015. "The Evolution of the Brain." Pp. 1961–1987 in *Handbook of Paleoanthropology*, edited by W. Henke and I. Tattersall. Berlin: Springer-Verlag.

Hopcroft, Rosemary. 2016. *Evolution and Gender: Why It Matters for Contemporary Life*. New York: Routledge.

Horowitz, Alan C. 2003. "Do Chimps Ape? Or Apes Human? Imitation and Intension in Humans (Homo sapiens) and Other Animals." *Journal of Comparative Psychology* 117: 325–336.

Huizinga, Johan. 1938 [1955]. *Homo Ludens: A Study of the Play-Element in Culture*. Boston: Beacon Press.

Hunt, K. D. and W. C. McGrew. 2002. "Chimpanzees in Dry Areas of Assirik, Snegal, and Semliki Wildlife Reserve, Uganda." Pp. 35–51 in *Great Apes Societies*, edited by W. C. McGrew, L. Marchant, and T. Nisida. Cambridge, UK: Cambridge University Press.

Itakura, S. 1996. "An Exploratory Study of Gaze-Monitoring in Non-Human Primates." *Japanese Psychological Research* 38: 174–180.

Jackson, Joshua C., Joseph Watts, Teague R. Henry, Johann-Mattis List, Robert Forkel, Peter J. Mucha, Simon J. Greenhill, Russel D. Gray, and Kristen A. Lindquist. 2019. "Emotion Semantics Show Both Cultural and Variation and Universal Structure." *Science* 366: 1217–1522.

Jeffers, Robert and Ilse Lechiste. 1979. *Principles for Historical Linguistics*. Cambridge, MA: M.I.T. Press.

Kanazawa, Satoshi. 2004. "The Savanna Principle." *Managerial and Decision Economics* 25: 41–54.

____. 2010. "Evolutionary Psychology and Intelligence Research." *American Psychologist* 65 (4): 279–289.

Kaneko, Talaaki and Masaki Tomonaga. 2011. "The Perception of Self-Agency in Chimpanzees (*Pan troglodytes*)." *Proceedings of the Royal Society* 278 (1725): 3694–3702.

Kelley, Jay. 2004. "Life History Evolution in Miocene and Extant Apes." Pp. 223–248 in *Human Evolution Through Developmental Change*, edited by Nancy Menugh-Purrus and Kenneth J. McNamara. Baltimore: Johns-Hopkins University Press.

Kelley, Jay and T. M. Smith. 2003. "Age at First Molar Emergence in Early Miocene *Afropithecus turkanensis* and Life-History Evolution in the *Hominoidea*." *Journal of Human Evolution* 44: 307–329.

Kemper, Theodore D. 1987. "How Many Emotions Are There? Wedding the Social and the Autonomic Components." *American Journal of Sociology* 93: 263–289.

Kemper, Theodore D. and Randall Collins. 1990. "Dimensions of Microinteraction." *American Journal of Sociology* 96: 32–68.

Kendrick, Douglas T., Jon K. Maner, and Norman P. Li. 2016. *Evolutionary Social Psychology*. Online publication in *The Handbook of Evolutionary Psychology*.

Kidder, A. V. 1940. "Looking Backward." *Proceedings of the American Philosophical Society* 83: 527–537.

Kitching, I. J., P. L. Forey, C. J. Humphries, and D. M. Williams. 1992 [1998]. *Cladistics 2nd Edition: The theory and Practice of Parsimony Analysis*. Oxford, UK: Oxford University Press.

Kluckhohn, Clyde. 1953. "Universal Categories of Culture." Pp. 507–523 in *Anthropology Today: An Encyclopedic Inventory*. Chicago: University of Chicago Press.

_____. 1959. "Common Humanity and Diverse Cultures." Pp. 245–284 in *The Human Meaning of The Social Sciences*, edited by Daniel Lerner. New York: Meridian.

Krutz, Gordon V. 1979. "The Native's Point of View as an Important Factor in Understanding the Dynamics of the Oraibi Split." *Ethnohistory* 20 (1): 77–89.

Kühl, Hjalmar S. et al. 2016. "Chimpanzee Accumulative Stone Throwing." *Nature Scientific Reports* 6 (22219): 1–8.

Lahnakoski, Juha M., Enrico Glerean, Iiro P. Jääskeläinen, Jukka Hyönä, Ritta Hari, Mikko Sams, and Lauri Nummenmaa. 2014. "Synchronous Brain Activity across Individuals Underlies Shared Psychological Perspectives." *NueroImage* 100: 316–324.

Langergraber, K. G. Schubert, C. Rowney, R. W. Wrangham, V. Reynolds, K. Hunt, and I. Vigilant. 2011. "Genetic Differentiation and the Evolution of Cooperation in Chimpanzees and Humans." *Proceedings of the Royal Society of London B* 278 (1717): 2546–2552.

Lawler, Edward J. 2001. "An Affect Theory of Social Exchange." *American Journal of Sociology* 107: 321–352.

Lawler, Edward J., Shane Thye, and Jeongkoo Yoon. 2009. *Social Commitments in a Depersonalized World*. New York: Russell Sage.

LeDoux, Joseph E. 1996. *The Emotional Brain: The Mysterious Underpinnings of Emotional Life*. New York: Simon and Schuster.

Lee, Richard B. 1979. *The !Kung San: Men, Women and Work in a Foraging Society*. Cambridge: Cambridge University Press.

Lee, Richard B. and Richard Daly, Eds. 1999. *The Cambridge Encyclopedia of Hunters and Gatherers*. Cambridge: Cambridge University Press.

Lenski, Gerhard. 1964. *Power and Privilege: A Theory of Stratification*. New York: McGraw-Hill.

_____. 2005. *Ecological-Evolutionary Theory: Principles and Applications*. New York: Roulteldge/Paradigm.

Lenski, Gerhard and Patrick Nolan. 2005. *Human Societies*. New York: McGraw-Hill.

Lents, Nathan H. 2016. *Not So Different: Finding Human Nature in Animals*. New York: Columbia University Press.

Lévi-Strauss, Claude. 1949 [1969]. *The Elementary Structures of Kinship*. Boston: Beacon Press.

Lewin, Kurt. 1948. *Resolving Social Conflicts*. New York: Harper and Brothers.

_____. 1951. *Field Theory in Social Science*. New York: Harper and Brothers.

Locke, John. 1689 [1988] *Two Treatises of Government*. Cambridge: Cambridge University Press.

Lopreato, Joseph. 1984. *Human Nature and Biocultural Evolution*. Boston MA: Allen and Unwin.

_____. 2001. "Sociobiological Theorizing: Evolutionary Sociology." Pp. 405–433 in *Handbook of Sociological Theory*, edited by J. H. Turner. New York: Kluwer Academic.

Lopreato, Joseph and Timothy Crippen. 1999. *Crisis in Sociology: The Need for Darwin*. London: Transaction.

Lukas, D., V. Reynolds, C. Goesch, and L. Vigilant. 2005. "To What Extent Does Living in a Group Mean Living with Kin?" *Molecular Ecology* 14: 2181–2196.

Lycett, Stephen J., Mark Collard, and William C. McGrew. 2010. "Are Behavioral Differences among Wild Chimpanzees Communities Genetic or Cultural? An Assessment of Using Tool-Use Data and Phylogenetic Methods." *American Journal of Physical Anthropology* 142: 461–467.

Maas, P. 1958. *Textual Criticism*. Oxford, UK: Oxford University Press.

MacSweeney, Mairéad, C. M. Capek, R. Campbell, B. Woll. 2008a. "The Signing Brain: The Neurobiology of Sign Language." *Trends in Cognitive Sciences* 11: 432–440.

MacSweeney, Mairéad, D. Waters, B. Woll, and U. Goswami. 2008b. "Phonological Processing in Deaf Signers and the Impact of Age of First Language Acquisition." *NeuroImage* 40: 1369–1379.

Malinowski, Bronisław. 1944 [1969]. *A Scientific Theory of Culture and Other Essays*. Oxford, UK: Oxford University Press.

Manica, Adrea, W. Amos, F. Balloux, and T. Hanihara. 2007. "The Effect of Ancient Population Bottlenecks on Human Phenotypic Variation." *Nature* 448: 346–348.

Marshall, Douglas A. 2017. "*Moral Origins of God: Darwin, Durkheim, and the Homo Duplex Theory of Theogenesis.*" *Frontiers of Evolutionary Sociology and Biosociology*. In press www.frontiersin.org.

Maryanski, Alexandra. 1986. "*African Ape Social Structure: A Comparative Analysis.*" Ph.D. Dissertation, University of California, Irvine.

_____. 1987. "African Ape Social Structure: Is There Strength in Weak Ties?" *Social Networks* 9: 191–215.

_____. 1992. "The Last Ancestor: An Ecological-Network Model on the Origins of Human Sociality." *Advances in Human Ecology* 2: 1–32.

_____. 1993. "The Elementary Forms of the First Proto-Human Society: An Ecological/ Social Network Approach." *Advances in Human Evolution* 2: 215–241.

_____. 1995. "African Ape Social Networks: A Blueprint for Reconstructing Early Hominid Social Structure." Pp. 67–90 in *Archaeology of Human Ancestry*, edited by J. Steele and S. Shennan. London: Routledge.

_____. 1996. "Was Speech an Evolutionary Afterthought?" Pp. 121–131 in *Communicating Meaning: The Evolution and Development of Language*, edited by B. Velichikovsky and D. Rumbaugh. Mahwah, NJ: Erlbaum.

_____. 1997. "The Origin of Speech and Its Implications for Optimal Size of Human Groups." *Critical Review* 12: 233–249.

_____. 2019. *Emile Durkheim and the Birth of the Gods.* New York and London: Routledge.

Maryanski, Alexandra, Peter Molnar, Ullica Segerstrale, and Borris M. Velichikovsky. 1997. "The Social and Biological Foundations of Human Communication." Pp. 181–200 in *Human by Nature*, edited by P. Weingart, S. Mitchell, P. Richerson, and S. Maasen. Mahwah, NJ: Erlbaum.

Maryanski, Alexandra, Stephen Sanderson, and Raymond Russell. 2012. "The Israeli Kibbutzim and the Westermarck Hypothesis." *American Journal of Sociology* 117: 1503–1508.

Maryanski, Alexandra and Jonathan H. Turner. 1992. *The Social Cage: Human Nature and the Evolution of Society.* Stanford, CA: Stanford University Press.

_____. 2018. "Incest: Theoretical Perspectives on." In *The International Encyclopedia of Anthropology*, edited by H. Callan. New York and London: John Wiley and Sons.

Maynard Smith, John. 1982. *Evolution and the Theory of Games.* Cambridge University Press.

McGrew, William. C. 1981. "The Female Chimpanzee as a Human Evolutionary Prototype." Pp. 35–73 in *Woman the Gatherer*, edited by Frances Dahlberg. New Haven: Yale University Press.

_____. 1983. "Animal Foods in the Diets of Wild Chimpanzees (*Pan troglodytes*): Why Cross Cultural Variation?" *Journal of Ethology* 1: 46–61.

_____. 1992. *Chimpanzee Material Culture: Implications for Human Evolution.* Cambridge, UK: Cambridge University Press.

_____. 2010. "In Search of the Last Common Ancestor: New Findings on Wild Chimpanzees." *Philosophical Transaction of the Royal Society* 365: 3265–3267.

McGrew, W. C., P. J. Baldwin, and G. E. G. Tutin. 1981. "Chimpanzees in Hot, Dry, and Open Habitat: Mt. Assirik, Senegal, West Africa." *Journal of Human Evolution* 10: 227–244.

McLennan, John F. 1896. *Studies in Ancient History.* London: Macmillan Co.

McNeill, William H. 1997. *Keeping Together in Time: Dance and Drill in Human History.* Cambridge, MA: Harvard University Press.

Mead, George Herbert. 1934. *Mind, Self, and Society.* Chicago, IL: University of Chicago Press.

_____. 1938. *Philosophy of the Act.* Chicago, IL: University of Chicago Press.

Meltzoff, A. N. 2002. "Imitation as a Mechanism of Social Cognition: Origins of Empathy, Theory of Mind, and the Representation of Action." Pp. 6–25 in *Handbook of Childhood Cognitive Development*, edited by U. Goswami. Oxford, UK: Blackwell Publishers.

Mendel, Gregor. 1866. "Versuche über Plflanzenhybriden." *Verhand- lungen des naturforschenden Vereines in Brünn, Bd. IV für das Jahr 1865*, Abhandlungen, 3–47.

Menzel, E. W. 1971. "Communication about the Environment in a Group of Young Chimpanzees." *Folia Primatologica* 15: 220–232.

Merker, Bjorn H., Guy S. Madison, and Patricia Eckerdal. 2009. "On the Role and Origin of Isochrony in Human Rhythmic Entrainment." *Cortex* 45 (1): 4–17.

Metzler, Dieter. 1991. "A. H. Anquetil-Duperron (1731–1805) und das Konzept der Achsenzeit." Pp. 123–133 in *Achaemenid History, VII*, edited by H. Sancisi-Weerdenburg and J. W. Drijvers. Leiden: Brill.

Mitani, J. C., D. A. Merriwether, and C. Zhang. 2000. "Male Affiliation, Cooperation and Kinship in Wild Chimpanzees." *Animal Behavior* 59: 885–893.

Mitani, J. C. and P. S. Rodman. 1979. "Territoriality: The Relation of Ranging Patterns and Home Range Size to Defendability, with an Analysis of Territoriality among Primate Species." *Behavioral Ecology and Sociobiology* 5: 541–551.

Mitani, J. C. and D. P. Watts. 2004. "Why Do Chimpanzees Share Meat?" *Animal Behavior* 61: 915–924.

Mitchell, P. 2011a. "Acquiring a Theory of Mind." In *An Introduction to Developmental Psychology*: Second Edition, edited by A. Slater and G. Bremner. BPS Blackwell.

———. 2011b. "Inferences about Guessing and Knowing by Chimpanzees (*Pan troglodytes*)." *Journal of Comparative Psychology* 104 (3): 203–210.

Moffett, Mark W. 2018. *The Human Swarm: How Our Societies Arise, Thrive, and Fall.* New York: Basic Books.

Moore, Deborah L., Kevin E. Bangergraber, and Linda Vigilant. 2015. Genetic Analyses Suggest Male Philopatry and Territoriality in Savanna-Woodland Chimpanzees (*Pan troglodytes schweinfurthii*) of Ugalla, Tanzania. *International Journal of Primatology*, 36: 377–397.

Morgan, Lewis Henry. 1871 [1997]. *Systems of Consanguinity and Affinitiy in the Human Family.* Lincoln, NE: University of Nebraska Press.

Morris, Desmond. 1967. *The Naked Ape: A Zoologist's Study of the Human Animal.* London: Jonathan Cape.

Mueller, Anna S. and Seth Abrutyn. 2016. "Adolescents under Pressure: A New Durkheimian Framework for Understanding Adolescent Suicide in a Cohesive Community." *American Sociological Review* 81 (5): 877–899.

Murdock, Peter. 1945. "The Common Denominator of Cultures." Pp. 123–142 in *The Science of Man in the World Crisis*, edited by Ralph Linton. New York: Columbia University Press.

Naroglwalla, M. C., Hutchison M. P., Begun, D. R. 2005. "Middle and Late Miocene Terrestrial Vertibrae Localities." *Beito. Palant.* 30: 319–31.

Newcombe, Theodore M. 1953. "An Approach to the Study of Communicative Acts." *Psychological Review* 60: 393–404.

Nolan, Patrick and Gerhard Lenski. 2018. *Human Societies.* New York: Oxford University Press.

Nummenmaa, Lauri, Enrico Glerean, Mikko Viinikainen, Iiro P. Jääskeläinen, Riitta Hari, and Mikko Sams. 2012. "Emotions Promote Social Interaction and Synchronizing Brain Activity across Individuals." *Proceedings of the National Academy of Sciences* 109 (24): 9599–9604.

Nummenmaa, Lauri, Heini Saarimäki, Enrico Glerean, Athanasios Gotsopoulos, Iiro P. Jääskeläinen, Riitta Hari, and Mikko Sams. 2014. "Emotional Speech Synchronizes Brains across Listeners and Engages Large-scale Dynamic Brain Networks." *NeuroImage* 102: 498–509.

Obhi, Sukhvinder S. and Natalie Sebanz. 2011. "Moving Together: Toward Understanding the Mechanisms of Joint Action." *Experimental Brain Research* 211: 329–336.

Okamoto, Sanae, Masaki Tomonaga, Kiyoshi Ishii, Nobuyuki Kawai, Masayuki Tanaka, and Tetsuro Matsuzawa. 2002. "An Infant Chimpanzee (*Pan troglodytes*) Follows Human Gaze." *Animal Cognition* 5: 107–114.

Olson, Mancur. 1967 [1971]. *The Logic of Collective Action*. Cambridge, MA: Harvard University Press.

Osgood, Charles E. 1966. "Dimensionality of the Semantic Space for Communication via Facial Expressions." *Scandinavian Journal of Psychology* 7: 1–30.

Panksepp, Jaak. 1982. "Toward a General Psychobiological Theory of Emotions." *Behavioral and Brain Sciences* 5: 407–467.

Parr, Lisa, Bridget Waller, and Jennifer Fugate. 2005. Emotional Communication in Primates: Implications for Neurology. *Current Opinion in Neurology*, 15, 716–720.

Passingham, Richard E. 1973. "Anatomical Differences between the Neocortex of Man and the Other Primates." *Brain Behavioral Evolution* 7: 337–359.

———. 1975. "Changes in the Size and Organisation of the Brain in Man and his Ancestors." *Brain and Behavior Evolution* 11: 73–90.

Piaget, Jean-Jacques. 1948. *The Moral Judgement of the Child*. Glencoe, IL: Free Press.

———. 1952. *The Origins of Intelligence in Children*. Glencoe, IL: Free Press.

———. 2002. *Emotions and Life: Perspectives from Psychology, Biology, and Evolution*. Washington, DC: American Psychological Association.

Plutchik, Robert 1962. *The Emotions: Facts, Theories, and a New Model*. New York: Random House.

———. 1980. *Emotion: A Psychoevolutionary Synthesis*. New York: Harper and Row.

———. 2002. *Emotions and Life: Perspectives from Psychology*. Washington, DC: American Psychological Association.

Povinelli, Daniel J. 2000. *Folk Physics for Apes: The Chimpanzee's Theory of How the World Works*. Oxford, UK: Oxford University Press.

———. 2003. "Chimpanzee Minds: Suspiciously Human?" *Trends in Cognitive Sciences* 7: 157–160.

Povinelli, D. J., K. E. Nelson, and S. T. Boysen. 1990. "Inferences about Guessing and Knowing by Chimpanzees (Pan troglodytes)." *Journal of Comparative Psychology* 104: 203–210.

Povinelli, Daniel J. and Timothy J. Eddy. 1997. "Specificity of Gaze-following in Young Chimpanzees." *British Journal of Developmental Psychology* 15: 213–222.

Povinelli, Daniel J. and Jennifer Vonk. 2003. "Chimpanzees Minds: Suspiciously Human?" Trends in Cognitivie *Sciences* 7: 157–160.

Premack, David and Guy Woodruff. 1978. "Does the Chimpanzee Have a Theory of Mind?" Behavior and Brain *Science* 1: 515–526.

Pruetz, J. D. and P. Bertolani. 2007. Savanna Chimpanzees Hunt with Tools. *Current Biology*, 17: 1–6.

Pruetz, Jill. 2006. "Feeding Ecology of Savanna Chimpanzees (Pan troglodytes verus)." Pp. 161–182 in *Feeding Ecology of Great Apes and Other Primates*, edited by G. Boesch, G. Hohmann, and M. Robbins. Cambridge, UK: Cambridge University Press.

Pruetz, Jill and Thomas C. LaDuke. 2010. "Reaction to Fire by Savanna Chimpanzees (*Pan troglodytes verus*) at Fongoli, Senegal." *American Journal of Physical Anthropology* 141: 646–650.

Reynolds, Vernon. 1965. *Budongo: An African Forest and Its Chimpanzees*. New York: Natural History Press.

Ridgeway, Cecilia. 1986. "Expectations, Legitimacy, and Dominance in Task Groups." *American Sociological Review* 51: 603–617.

_____. 2001. "Inequality, Status, and the Construction of Status Beliefs." In *Handbook of Sociological Theory*, edited by J. H. Turner. New York: Plenum Scientific.

Ridgeway, Cecilia and K. G. Erickson. 2000. "Creating and Spreading Status Beliefs." *American Journal of Sociology* 106: 579–615.

Rizzolattti, Giacomo, Luciano Fadiga, Leonardo Fogassi, and Vittorio Gallese. 2002. "From Mirror Neurons to Imitation: Facts and Speculations." Pp. 247–266 in *The Imitative Mind: Development, Evolution and Brain Bases*, edited by W. Prinz and A. N. Meltzoff. Cambridge, UK: Cambridge University Press.

Rizzolatti, Giacomo and Gorrado Sinigalia. 2008. *Mirrors in the Brain: How Our Minds Share Actions, Emotions, and Experience*. Oxford, UK: Oxford University Press.

Rousseau, Jean-Jacques. 1762 [1968]. *Du contrat social; ou, Principes du droit politique*. New York: Penguin Book translation.

Rumbaugh, M. Duane. 2013. *With Apes in Mind: Emergence, Communication and Competence*. Distributed by Amazon.com.

_____. 2015. "A Salience Theory of Learning and Behavior and Rights of Apes." Pp. 514–536 in *Handbook on Evolution and Society: Toward an Evolutionary Social Science*, edited by J. H. Turner, R. Machalek, and A. Maryanski. New York: Routledge/Paradigm.

Rumbaugh, Duane M. and D. A. Washburn. 2003. *Intelligence of Apes and Other Rational Beings*. New Haven CT: Yale University Press.

Rumbaugh, Duane and E. Sue Savage-Rumbaugh. 1990. "Chimpanzees: Competencies for Language and Numbers." Pp. 409–441 in *Comparative Perception: Complex Signals*, Volume 2, edited by W. Stebbins and M. Berkley. New York: Wiley and Sons.

Sachs, Harvey. 1972. "An Initial Investigation of the Usability of Conversational Data for Doing Sociology." Pp. 31–74 in *Studies on Conversation*, 2 Volumes, edited by D. Sudnow. New York: Free Press.

Sachs, Harvey, Emanuel Schegloff, and Gail Jefferson. 1974. "A Simple Systematics for the Analysis of Turn Taking in Conversation." *Language* 50: 696–697.

Sanderson, Stephen K. 2001. *The Evolution of Human Sociality. A Darwinian Conflict Perspective*. Lanham, MD: Rowman and Littlefield.

Savage-Rumbaugh, Sue and Roger Lewin. 1994. *Kanzi: The Ape at the Brink of the Human Mind*. New York: John Wiley and Sons.

Savage-Rumbaugh, Sue, Rose A. Seveik, and William D. Hopkins. 1988. "Symbolic Cross-Model Transfer in Two Species." *Child Development* 59: 617–625.

Savage-Rumbaugh, Sue E., Jeannine Murphy, Rose A. Seveik, Karen E. Brakke, Shelly L. Williams, and Duane M. Rumbaugh. 1993. *Language Comprehension in the Ape and Child (Monographs of the Society for Research in Child Development 58)*. Chicago: University of Chicago Press.

Savage-Rumbaugh, Sue E., Duane Rumbaugh, and Sally Boysen. 1978. "Symbolic Communication Between Two Chimpanzees." *Science, New Series* 201 (4356): 641–644.

Saxe, Rebecca and N. Kanwisher. 2003. "People Thinking about Thinking People: The Role of the Temporo-Parietal Junction in 'Theory of Mind.'" *NeuroImage* 19: 1835–1842.

Saxe, Rebecca, Laura E. Schulz, and Yuhong V. Jiang. 2006. "Reading Minds Versus Following Rules: Dissociating Theory of Mind and Executive Control in the Brain." *Social Neuroscience* 1: 284–298.

Schaller, George. 1962. *The Ecology and Behavior of the Mountain Gorilla*. Ph.D. dissertation, University of Wisconsin.

Scheff, Thomas. 1988. "Shame and Conformity: The Deference-Emotion System." *American Sociological Review* 5: 395–406.

Schulze, Katrin, Faraneh Varga-Kadem, and Mortimer Mishkin. 2018. "Phonological Working Memory and FOXP." *Neuropsychologia* 108: 147–152.

Schutz, Alfred. 1932 [1967]. *The Phenomenology of the Social World*. Evanston, IL: Northwestern University Press.

Semendeferi, Katerina and Hanna Damasio. 2000. "The Brain and Its Main Anatomical Subdivisions in Living Hominoids Using Magnetic Resonance Imaging." *Journal of Human Evolution* 38: 317–332.

Semendeferi, Katerina, A. Lu, Natalie Schenker, and Hanna Damasio. 2002. "Humans and Great Apes Share a Large Frontal Cortex." *Nature Neuroscience* 5: 272–276.

Seneviratne, H. L. 1999. *The Work of Kings: The New Buddhism in Sri Lanka*. Chicago: University of Chicago Press.

Sherwood, Chet C. 2007. "The Evolution of Neuron Types and Cortical Histology in Apes and Humans." Pp. 355–378 in *Evolution of Nervous Systems 4: The Evolution of Primate Nervous Systems*, edited by T. M. Preuss and J. H. Kaas. Oxford, UK: Academic Press.

Sherwood, Chet C., Ralph L. Holloway, Katerina Semendeferi, and Patrick R. Hof. 2005. "Is Prefrontal White Matter Enlargement a Human Evolutionary Specialization?" *Nature Neuroscience* 8: 537–538.

Sherwood, Chet C., Francys Subiaul, and Tadeusz W. Zawidzki. 2008. "A Natural History of the Human Mind: Tracing Evolutionary Changes in Brain and Cognition." *Journal of Anatomy* 212: 426–454.

Simmel, Georg. 1906 [1990]. *The Philosophy of Money*. Translated T. Bottomore and D. Frisby. Boston, MA: Routledge.

Smith, Adam. 1776. *An Inquiry into the Nature of Causes of The Wealth of Nations*. Shine Classics. *amazon.com/books*.

Smith, Christian. 2010. *What Is a Person?* Chicago, IL: University of Chicago Press.

Smith, Lindsey W. and Roberto A. Delgado. 2015. "Body Language: The Interplay between Positional Behavior and Gestural Signaling in the Genus *Pan* and Its Implications for Language Evolution." *American Journal of Physical Anthropology* 157: 592–602.

Spencer, Herbert. 1874–96 [1899]. *The Principles of Sociology 3 volumes*. New York: Appleton-Century-Crofts.

Spocter, Muhammad A., William D. Hopkins, Amy R. Garrison, Amy L. Bauernfeind, Cheryl D. Stimpson, Patrick R. Hof, and Chet C. Sherwood. 2010. "*Wernicke's Area Homologue in Chimpanzees (Pan troglodytes) and Its Relation to the Appearance of Modern Human Language*." *Proceedings of the Royal Society B*. doi:10.1098/rspb.2010.0011: 1–10.

Stanford, C. B. 1990. *The Hunting Apes: Meat Eating and the Origins of Human Behavior*. Princeton, NJ: Princeton University Press.

Stebbins, G. Ledyard. 1969. *The Basis of Progressive Evolution*. Chapel Hill, NC: University of North Carolina Press.

Stephan, Heinz. 1983. "Evolutionary Trends in Limbic Structures." *Neuroscience and Biobehavioral Reviews* 7: 367–374.

Stephan, Heinz and O. J. Andy. 1969. "Quantitative Comparative Neuroanatomy of Primates: An Attempt at Phylogenetic Interpretation." *Annals of the New York Academy of Science* 167: 370–387.

_____. 1977. "Quantitative Comparison of the Amygdala in Insectivores and Primates." *Acta Antomica* 98: 130–153.

Stephan, Heinz, Georg Baron, and Heiko Frahm. 1986. "Comparative Size of Brains and Brain Components." Pp. 1–37 in *Comparative Primate Biology*, Volume 4, edited by H. Steklis and J. Erwin. New York: Alan R. Liss.

Stephen, Heinz, Heiko Frahm, and Georg Baron. 1981. "New and Revised Data on Volumes of Brain Structures in Insectivores and Primates." *Folia Primatoligica* 35: 1–29.

Stets, Jan E. and Richard T. Serpe, Eds. 2016. *New Directions in Identity Theory and Research*. New York: Oxford University Press.

Subiaul, F. 2007. "The Imitation Faculty in Monkeys: Evaluating Its Features, Distribution, and Evolution." *Journal of Anthropological Science* 85, 35–62.

Symons, Donald. 1979. *The Evolution of Human Sexuality*. New York: Oxford University Press.

Taglialatela, J. P., J. L. Russell, J. A. Schaeffer, and W. D. Hopkins. 2008. "Communicative Signaling Activates 'Broca's' Homolog in Chimpanzees." *Current Biology* 18: 343–348.

Tiger, Lionel and Robin Fox. 1971. *The Imperial Animal*. New York: Holt, Rinehart, and Winston.

Tomasello, Michael, Brian Hare, and Tara Fogleman. 2001. "The Ontogeny of Gaze Following in Chimpanzees, *Pan troglodytes*, and Rhesus Macaques, *Macaca mulatta*." *Animal Behavior* 61: 335–343.

Tomonaga, Michael. 1999. "Attending to the Others' Attention in Macaques' Joint Attention or Not?" *Primate Research* 15: 425.

Tooby, John and Leda Cosmides. 1992. "The Psychological Foundations of Culture." Pp. 19–136 in *The Adapted Mind: Evolutionary Psychology and the Generation of Culture*, edited by Jerome H. Barkow, Leda Cosmides, and John Tooby. New York: Oxford University Press.

Trivers, Robert L. 1971. "The Evolution of Reciprocal Altruism." *Quarterly Review of Biology* 46: 35–57.

_____. 1972. "Parental Investment and Sexual Selection." Pp. 136–179 in *Sexual Selection and the Descent of Man*, edited by B. G. Campbell. Chicago IL: Aldine.

_____. 1974. Parent-Offspring Conflict. *American Zoologist*, 14: 249–264.

_____. 2004. "Parent-Offspring Conflict." *American Zoologist* 14: 249–264.

_____. 2005. "Reciprocal Altruism: 30 Years Later." Pp. 67–83 in *Cooperation in Primates and Humans: Mechanisms of Evolution*, edited by P. M. Kappeler and C. P. van Schaik. New York: Springer.

Turner, Jonathan. 1987. "Toward a Sociological Theory of Motivation." *American Sociological Review* 52: 15–27.

_____. 1988. *A Theory of Social Interaction*. Stanford, CA: Stanford University Press.

_____. 1995. *Macrodynamics: Toward a Theory on the Organization of Human Populations*. New Brunswick, NJ: Rutgers University Press.

_____. 1996a. "The Evolution of Emotions in Humans: A Darwinian-Durkheimian Analysis." *Journal for the Theory of Social Behaviour* 26: 1–34.

_____. 1996b. "Cognition, Emotion, and Interaction in the Big-Brained Primate." *Contemporary Studies in Sociology* 13: 297–318.

_____. 1996c. "Toward a General Sociological Theory of Emotions." *Journal for the Theory of Social Behavior* 29: 132–162.

_____. 1997a. *The Institutional Order*. New York: Longman.

_____. 1997b. "The Evolution of Emotions: The Nonverbal Basis of Human Social Organization." Pp. 211–228 in *Nonverbal Communication: Where Nature Meets Culture*, edited by Ullica Segerstrale and Peter Molnar. Hillsdale, NJ: Erlbaum.

_____. 1998. "The Evolution of Moral Systems." *Critical Review* 11: 211–232.

_____. 1999. "The Neurology of Emotions: Implications for Sociological Theories of Interpersonal Behavior." *Social Perspectives on Emotion* 5: 81–108.

_____. 2000. *On the Origins of Human Emotions: A Sociological Inquiry into the Evolution of Human Affect*. Stanford, CA: Stanford University Press.

_____. 2002. *Face to Face: Toward a Theory of Interpersonal Behavior*. Palo Alto: Stanford University Press.

_____. 2003. *Human Institutions: A New Theory of Societal Evolution*. Lanham, MD: Rowman and Littlefield.

_____. 2007. *Human Emotions: A Sociological Theory*. Oxford, UK: Routledge.

_____. 2010. *Theoretical Principles of Sociology*, Volume 2 on Microdynamics. New York: Springer.

_____. 2011. "Extending the Symbolic Interactionist Theory of Interaction Processes: A Conceptual Outline." *Symbolic Interaction*, 34(3): 330–339.

_____. 2013. *Theoretical Principles of Sociology*, Volume 2 on Mesodynamics. New York: Springer.

_____. 2019. "The Effects of Cultural, Structural, and Interpersonal Dynamics on Interaction Rituals." *Essays in Honor of Randall Collins*, New York and London: Routledge 2019.

_____. 2021. "Why Are Humans So Emotional: An Explanation from Evolutionary Sociology." In *Oxford Handbook on Emotional Development*, edited by D. Dukes, A. Samson, and E. Walle, Eds., in press.

Turner, Jonathan H. 1984. *Societal Stratification: A Theoretical Analysis*. New York: Columbia University Press.

Turner, Jonathan H. and Richard Machalek. 2016. *The New Evolutionary Sociology*. New York: Routledge.

Turner, Jonathan H., Richard Machalek, and Alexandra Maryanski, Eds. 2015. *Handbook of Evolution and Society: Toward an Evolutionary Social Science*. New York: Paradigm.

Turner, Jonathan H. and Alexandra Maryanski. 2005. *Incest: Origins of the Tabooized Theoretical and Methodologica*. New York and London: Routledge/Paradigm.

_____. 2008. *On The Origins of Societies by Natural Selection*. Boulder, CO: Paradigm Press.

_____. 2015. "Evolutionary Sociology: A Cross-Species Strategy for Discovering Human Nature." Pp. 546–571 in *Handbook of Evolution and Society: Toward an Evolutionary Social Science*, edited by J. H. Turner, R. Machalek, and A. R. Maryanski. Boulder, CO: Paradigm/New York: Routledge.

_____. 2019. "Emotions and The Evolution of Human Auditory Language: An Application of Evolution and Neuro Sociology." In *The Handbook of Language and Emotions*, edited by J. Fenigsen, S. Pritzker, and J. Wilce. New York and London: Routledge.

_____. 2021. Emotions and the Evolution of Auditory Language: An Application of Neuro Sociology. In *The Handbook of Language and Emotions*, edited by J. Fenigisen, S. Pritzker, and J. Wilce. New York and London: Routledge.

Turner, Jonathan H., Alexandra Maryanski, Anders Peterson, and Armin Geertz. 2018. *The Emergence and Evolution of Religion: By Means of Natural Selection.* New York and London: Routledge.

Turner, Jonathan H. and Jan E. Stets. 2005. *The Sociology of Emotions.* New York: Cambridge University Press.

Turner, Ralph H. 1962. "Role-taking: Processes versus Conformity." Pp. 20–40 in *Human Behavior and Social Processes,* edited by A. Rose. Boston: Houghton Mifflin.

Tutin, W., W. C. McGrew, and P. J. Baldwin. 1982. "Responses of Wild Chimpanzees to Potential Predators." Pp. 136–141 in *Primate Behavior and Sociobiology,* edited by B. Chiarelli and R. Corruccini. Berlin: Springer-Verlag.

van den Berghe, Pierre. 1972. *Intergroup Relations: Sociological Perspectives.* New York: Elsevier.

____. 1973. *Age and Sex in Human Society.* Belmont, CA: Wadsworth.

____. 1975. *Man in society: A Biosocial View.* New York: Elsevier.

____. 1981. *The Ethnic Phenomenon.* New York: Elsevier.

de Waal, Frans B. M. 1982. *Chimpanzee Politics.* Baltimore: Johns-Hopkins University Press.

____. 1989. "Food Sharing and Reciprocal Obligations among Chimpanzees." *Journal of Human Evolution* 18: 433–459.

____. 1991. "The Chimpanzee's Sense of Social Regularity and Its Relation to the Human Sense of Justice." *American Behavioral Scientist* 34: 335–349.

____. 1996. *Good Natured: The Origins of Right and Wrong in Humans and Other Animals.* Cambridge, MA: Harvard University Press.

____. 2009. *The Age of Empathy: Nature's Lessons for a Kinder Society.* New York: Three Rivers Press.

____. 2016. *Are We Smart Enough to Know How Smart Other Animals Are?* New York: W. W. Norton.

____. 2019. *Mama's Last Hug: Animal Emotions and What They Tell Us about Ourselves.* New York: W. W. Norton.

de Waal, Frans B. M. and Sarah F. Brosnan. 2006. "Simple and Complex Reciprocity in Primates." Pp. 85–106 in *Cooperation in Primates and Humans: Mechanisms and Evolution,* edited by P. Kappeler and C. P. van Schaik. Berlin: Springer-Verlag.

Watts, D. P. and J. C. Mitani. 2001. "Boundary Patrols and Intergroup Encounters in Wild Chimpanzees." *Behaviour* 138: 299–327.

Weiner, Bernard. 1986. *An Attribution Theory of Motivation and Emotion.* New York: Springer.

Wells, Spencer. 2002. *The Journey of Man: A Genetic Odyssey.* Princeton, NJ: Princeton University Press.

Westermarck, Edmund 1891. *The History of Marriage.* London: Macmillan.

____. 1981 [1922]. *The History of Marriage, three volumes.* New York: Appleton.

____. 1926. *A Short History of Marriage.* London: Macmillan.

Whitehead, Hal and Luke Rendell. 2015. *The Cultural Lives of Whales and Dolphins.* Chicago IL: University of Chicago Press.

Whitely, Peter M. 1988. *Deliberate Acts: Changing Hopi Culture through the Oraibi Split.* Tucson: University of Arizona Press.

____. 2008. *The Orayvi Split: A Hopi Transformation. Vol. 1: Structure and History. Vol. 2: The Documentary Record.* New York: American Museum of Natural History.

Williams, George C. 1966. *Adaptation and Natural Selection. A Critique of Some Evolutionary Thought.* Princeton, NJ: Princeton University Press.

Wilson, David Sloan. 1975. "A Theory of Group Selection." *Proceedings of the National Academy of Sciences* 72: 143–146.

_____. 2002. *Darwin's Cathedral. Evolution, Religion, and the Nature of Society.* Chicago University Press.

Wilson, David Sloan and Edward O. Wilson. 2007. "Rethinking the Theoretical Foundations of Sociobiology." *The Quarterly Review of Biology* 82 (4): 327–348.

Wilson, Edward O. 1975. *Sociobiology: The New Synthesis.* Cambridge, MA: Harvard University Press.

_____. 1978. *On Human Nature.* Cambridge, MA: Harvard University Press.

_____. 1998. *Consilience: The Unity of Knowledge.* London: Little, Brown.

_____. 2019. *Genesis: The Deep Origins of Society.* New York:: W. W. Norton.

Winkelman, Michael. 1998. "Aztec Human Sacrifice: Cross-Cultural Assessments of the Ecological Hypothesis." *Ethnology* 37 (3): 285–298.

Wohlgemuth, Sandra, Iris Adam, and Constance Scharf. 2014. "FoxP2 in Songbirds." *Current Opinion in Neurobiology* 28: 86–93.

Wolpoff, Milford H. 1999. *Paleoanthropology.* Boston: McGraw-Hill.

Wuthnow, Robert. 1987. *Meaning and the Moral Order: Explorations in Cultural Analysis.* Berkeley, CA: University of California Press.

Wynne-Edwards, V. C. 1962. *Animal Dispersion in Relation to Social Behavior.* Edinburgh, Scotland: Oliver and Boyd.

_____. 1986. *Evolution through Group Selection.* Oxford, UK: Blackwell Scientific.

Zollikofer, Christoph P. E. and Marcia Silva Ponce de Leon. 2013. "Pandora's Growing Box: Inferring the Evolution and Development of Hominin Brains from Endocasts." *Evolutionary Anthropology* 22: 20–33.

Subject Index

exchange dynamics within, 235–37
Goffman's conception of, 231
longer-term rituals, 231–32
rhythmic synchronization of, 105–106
shorter-term rituals, 231
totemizing rituals, 233–35
Interpersonal "fronts" 215

Justice
among monkeys and apes, 111
calculations for, 111
and emotional arousal, 111–12
normative basis of, 112
and social solidarity, 112

Kanzi
and emphasizing, 104
and language use, 103
Keying and Rekeying, of frames, 152–53
Kin selection 12–13
problems with, 183
Trivers' view, 12

Lamination of frames 229
Language
and culture, 75
and emotions, 64–67
of emotions, 67–73, 101
and inferior parietal lobe, 70
pre-adaptation, 45
Last Common Ancestor (LCA) 28,
35, 37
Life history characteristics 45–46,
85–89
Looking glass self 109

Macro-societies, pre-adaptations
for, 245–56
Mate selection 4
Memes 11
Memory
and emotions, 57–63
and frontal lobe, 62–63
and hippocampus, 63
Mind, J. Dewey's conception, 139–40
Mirror neurons 101
Mirror test for self 108

Modern synthesis in biology 8, 54
Modules, of the brain, 13, 15
Mutations 54–55
Need-states
for congruence/consistency, 196–98
for efficacy, 193–95
for group inclusion, 195–96
for identity verification, 184–92
for positive emotions, 199–201
for profitable exchange payoffs, 192–93
and psychology complex, 204–205
for trust, 198–99
Negative emotions, and exchange, 192
Neuroanatomy, selection on, 55
Nuclear family, evolution of, 88, 93–96

Ordering mechanisms
abstraction, 151–52
attributions, 157–58
categorization, 154–55
chunking, 152–53
cognitive congruence, 155
contrast-conceptions, 155–56
emotions and memory, 148–50
expectations states, 156–57
framing, 152
Gestalt dynamics, 154
hand, 149–51
in human brains, 148–152
response generalization 151–52
salience, 151–52
stocks of knowledge at,
Over-population 262–63

Parental investment 14
Phenotypes 8
Pre-adaptations
definition of, 44
list of, 45
Prefrontal cortex 60–63
Play 45, 91–92
among great apes, 91–92
as pre-adaptation, 45
Population declines, early humans,
253–55
Projection 79
Psychology complex, the, 201–206

Name Index